Guide to German Seaplanes of WWI

Jack Herris

Great War Aviation Centennial Series #58

This book is dedicated to readers trying to identify early German seaplanes.

Acknowledgements
Thanks to Mike O'Neal for the cover painting.

For our aviation books in print please see our website at: **www.aeronautbooks.com**.
I am looking for photographs of the less well-known German aircraft of WWI for future titles. For questions or to help with photographs please contact me at **jherris@me.com**

ISBN: 978-1-953201-04-1

Design and layout: Jack Herris
Cover design: Aaron Weaver
Digital photo editing: Jack Herris

Aeronaut Books
Books for Enthusiasts by Enthusiasts
www.aeronautbooks.com

Table of Contents

Preface

While writing this series on WWI German warplanes, it was often difficult to identify the multitude of German seaplanes. The seaplanes are generally more differentiated than the B-types, but many are not well known. The Navy had a great variety of different seaplane types, often prototypes or production seaplanes built in small numbers.

Fortunately, there is a nearly complete list of Marine Numbers assigned to different seaplanes that gives the type for each Marine Number. This essential list is reproduced in this book starting on page 5.

Friedrichshafen built more seaplanes for the German Navy than any other company, followed by Gotha and Hansa-Brandenburg. Many other companies built prototype floatplanes and a few were placed in production. The largest production of any single floatplane was 118 Albatros W.4 single-seat fighters in seven production batches plus the prototypes.

The purpose of this book is to illustrate the different German seaplanes and identifying their distinguishing characteristics to help readers identify the types. Readers wanting more information, including color profiles, scale drawings, specifications, production data, or more photos, are directed to the earlier titles in the series.

Jack Herris

Left: Friedrichshafen FF33E Marine Number 738 operating with a U-boat; it was also a class HFT ordered in May 1916. The FF33E was the workhorse of German naval aviation for most of the war; gradually it was replaced by the similar but more powerful FF49.

Above: W.29 monoplanes on patrol over the North Sea in late summer 1918 with a W.19 of greater endurance.

Below: The Gotha UWD shows more of its distinctive profile in this dramatic image.

German Seaplane Numbers

M/Nr. #	Type	Class	Engine
E1	Rumpler Taube		100 Argus
D2	Albatros		70 Merc.
D3	Albatros WMZ		
E4	Rumpler Taube		100 Argus
D5	Albatros WMZ		75 Nag
D6	Curtis		
D7	Ago		100 Argus
E8	Rumpler Taube		100 Argus
D9	Albatros WMZ		As.I
D10	Albatros WMZ		As.I
D11	Albatros WMZ		As.I
D12	Avro		100 Gnome
13	Wight		Salmson
14	Albatros		D.I
D15	Ago		145 Argus
D16	Albatros		As.II
D17	Albatros WMZ		Bz.I
D18	Curtis		90 Curtiss
D19	Ago		100 Gnome
20	Albatros	FT	D.I
21	Albatros		D.I
22	Albatros		D.I
23	Albatros		As.II
24	Albatros		As.II
25–29	FF19		D.I
30–39	Ago		100 Gnome
40	Albatros		D.I
41	FF11		135 Nag
42	Lohner Type M		D.I
43	Curtis		
44	Sopwith		200 Salmson
45	AEG		Bz.III
46	Oertz FB3		D.III
47	Rumpler 4E		120 AD
48	Aviatik WP18		200 hp Argus
49	Rumpler 4B11		D.I
50	Rumpler 4B13		160 Gnome
51	Rumpler 4B12		Bz.III
52	Albatros		Bz.III
53	Albatros WDD	B	D.III
54	Albatros WDD		200 Argus
55	Albatros WDD	B	D.III
56	Albatros WDD		Bz.III
57	Brandenburg	B	Bz.III
58	Brandenburg	B/S	D.III
59	Gotha WD1		D.I
60	Gotha WD2		Bz.III
61	Gotha WD2		Rp.III
62	FF27		135 Nag
63	Oertz W4		D.III
64	Rumpler		D.I
65–69	Ago		As.III

M/Nr. #	Type	Class	Engine
70	Ago		As.III
71	Ago		As.III
72	Ago		U.III
71–72	Brandenburg W		Mb.III
Numbers 71–72 re-assigned to Brandenburg trainers			
73	Union		120 AD
74	Albatros		Bz.III
75	Oertz FB2		As.II
76–85	FF19	B	D.I
86–90	Rumpler 4B11		D.I
91–98	FF29A	B	D.I
99–100	FF33A	B	D.I
101–110	Rumpler 4B12	B	Bz.III
111	Albatros		D.I
112	Ago		100 Ob
113	Ago		100 Ob
114	Ago		100 Ob
115	Ago C.Iw	CB	Bz.III
116	Lohner (Br.AE)		Bz.III
117	FF34	CHFT	Mb.IV
118	Gotha WD5		D.III
119	Gotha WD7		2xD.II
120	Gotha UWD		SxD.III
121–186	Open		
187	Curtis		
188–200	Open		
201–210	FF29	B	D.II
211	FF33	B	D.II
212	FF29		D.II
213–220	FF33	B	D.II
221–230	Albatros	B	Bz.III
231–235	Brandenburg W	BFT	Bz.III
236–240	Gotha WD2	B	D.III
241–253	Rumpler 4B12	B	Bz.III
254–258	Gotha WD2		D.III
259	Gotha WD3	BMG	2xD.III
260–273	Brandenburg W	B	Bz.III
274–275	FF31	CB	Mb.III
276–280	Oertz W5	CB	Mb.IV
281	Oertz W6		2xMb.IV
282–284	Trav. F1	B	D.III
285–289	Gotha WD1	B	D.I
290	FF29	B	As.II
291–296	FF33D	B	As.II
297–299	Open		
300	FF35	G	2xD.III
301–400	Open		
401	KW Whlm. W1	B	D.I
402	KW Whlm. W2	B	D.I
403	KW Whlm. W3	B	D.I
404	KW Danzig		D.I
405	KW Danzig	S	D.I

Mr. #	Type	Class	Engine
406–410	FF29	B	D.II
411–415	FF29A	B	D.I
416–421	FF33A	B	D.II
422–423	Brandenburg W	B	Bz.III
424–425	Gotha WD2	C	D.III
426	FF33	B	D.II
427–430	Open		
431	Albatros	B	Bz.III
432	Albatros	B	D.I
433–435	Albatros		Cancelled
436–442	Rumpler 4B12	B	Bz.III
443–445	LVG	B	D.III
446–447	Albatros	B	Bz.III
448–449	Open		
450	Albatros W2	C	D.III
451–455	FF33	B	D.II
456–458	FF33B	BFT	Mb.III
459–460	FF33D	CB	Mb.III
461	KW Whlm. W	B	Bz.III
462	KW Whlm. W6	B	Bz.III
463–466	KW Kiel	B	Bz.III
467–470	KW Danzig	B	Bz.III
471–473	FF33C	BFT	D.III
474	Oertz W7	C	Mb.III
475	Oertz W7	C	D.III
476	Gotha WD8	C	Mb.IV
477	Brandenburg LW	CBMG	D.III
478–483	FF33E	BFT	Bz.III
484	Lohner T	C	Bz.III
485	Brandenburg LW	B	D.III
486–489	Brandenburg NW	B	D.III
490	Sablatnig SF1	B	D.III
491–496	FF33E	BFT	Bz.III
497–508	FF33E	B	Bz.III
509	LFG Roland W1	B	Bz.III
510	FF33H	C	Bz.III
511–516	Brandenburg FB	C	Bz.III
517–526	Brandenburg NW	B	D.III
527	Albatros W3	TMG	2xBz.III
528	Brandenburg GW	TMG	2xD.III
529–533	FF33E	B	Bz.III
534–538	FF33F	B	Bz.III
539–540	Ago C.IIw	C	Bz.III
541–550	FF33E	BFT	Bz.III
551	Ru- KW Danzig	B	Bz.III
552	Brand. Danzig	B	Bz.III
553–562	FF33E	BFT	Bz.III
563–570	Brandenburg NW	BFT	D.III
571	Brandenburg LW	C	D.III
572	Gotha WD9	C	D.III
573–577	Gotha WD2	CB	Bz.III
578	FBA		
579	Grigorovitch		100 hp Mono.
580–585	Sablatnig SF2	BFT	D.III

Mr. #	Type	Class	Engine
586	Ago C.IIw	C	Bz.IV
587	FF39	CHFT	Bz.IV
588–590	Brand.KW	CFT/S	Bz.IV
591–595	FF33E	BFT	Bz.III
596–600	FF33H	C	Bz.III
601–608	Brand.NW	B	D.III
609–618	Sablatnig SF2	B	D.III
619	Sablatnig SF3	CHFT	Bz.IV
620–624	Brand. GW	TMG	2xD.III
625–632	Brand. GNW	B	D.III
633–637	FF33E	BFT	Bz.III
638–642	FF33H	C	Bz.III
643–645	FF39	CHFT	Bz.IV
646–650	Brand. GW	TMG	2xD.III
651–658	Brand. GNW	BFT	D.III
659–663	FF33E	BFT	Bz.III
664–668	FF33H	C	Bz.III
669	FF40	CFT	Mb.IV
670–676	Gotha WD7	G	2xD.I
677	Travemünde F2	CFT	D.IV
678	FF 41A	TMG	2xBz.III
679	Gotha WD11	TMG	2xD.III
680–684	FF33E	BFT	2xBz.III
685–689	FF33H	C	Bz.III
690–694	FF33E	BFT	Bz.III
695–699	FF33H	C	Bz.III
700–704	Brand. GW	TMG	2xD.III
705–714	Sablatnig SF2	BFT	D.III
715–724	FF33E	BFT	Bz.III
725–729	FF33E	HFT	Bz.III
730–734	FF33H	C	Bz.III
735	FF	T	2xBz.IV
736–740	FF33E	BFT	Bz.III
741–745	FF33H	C	Bz.III
746	Brand. GDW	T/G	2xBz.IV
747	Albatros W4	ED	D.III
748	Brand. KDW	ED	Bz.III
749	FF43	ED	D.III
750	LFG W	ED	D.III
751	Rumpler 6B1	ED	D.III
752–781	Brand. NW(Go)	B	D.III
782	Ursinus-Rex	ED	150 Bz.III
783–784	Brand. KDW	ED	150 Bz.III
785–786	Albatros W4	ED2MG	150 Bz.III
787–788	Rumpler 6B1	ED	150 Bz.III
789–790	LFG W	ED	Cancelled
789–790	FF33S(Rol)	S	D.III
791–800	Sablat. SF2(LVG)	B	D.III
801	Gotha WD14	TMG	2xBz.IV
802–811	FF33E	B	Bz.III
812–821	FF33H	C	Bz.III
822–841	FF33E	HFT	Bz.III
841	FF33E *Wölfchen* replacement for propoganda		
842–843	Gotha WD15	B	D.IVa

Mr. #	Type	Class	Engine
844	Travemünde F3-Refused	ED	Bz.III
845–849	Albatros W5	TMG	2xBz.III
850–859	FF33E	HFT	Bz.III
860–869	FF33E	B	Bz.III
870–884	FF33E	HFT	Bz.III
885–889	FF33E	B	Bz.III
890–899	Rumpler 6B1	ED	D.III
900	Sab. SF4(LVG)	ED	Bz.III
901	Sab. SF4	ED	Bz.III
902–911	Albatros W4	ED	D.III
912–921	Brand. KDW	ED	Mb.III
922–931	FF33J	HFT	Bz.III
932	FF33K	C2MG	Bz.III
933–941	FF33L	HFT	Bz.III
942	Roland	BFT	D.III
943	Roland W16	CFT	Bz.IV
944	Gotha WD12A	BFT	D.III
945	KW Whlm. W9	C2MG	Bz.III
946	Brand. CC	E	Bz.III
947	KW Whlm.	CHFT	D.IV
948–967	Albatros W4	E2MG	D.III
968–975	Sablatnig SF5	B	Bz.III
976–987	Sablatnig SF5	HFT	Bz.III
976 (Ersatz)	Repl. original 976	HFT	Bz.III
988–990	Brand. W11	ED	Bz.IV
991–995	Gotha WD11	TMG	2xD.III
996–1000	FF41A	TMG	2xBz.III
1001–1010	FF33L	CHFT	Bz.III
1011–1016	Brand. W12	C2MG	D.III
1017–1036	Sablat. SF5(LVG)	B	Bz.III
1037–1061	Rumpler 6B1	ED2MG	D.III
1062–1066	Rumpler 6B2	ED2MG	D.III
1067–1076	Brand. KDW	ED	Bz.III
1077–1079	Brand. W16	ED2MG	U.III
1080–1084	Brand. GW	T/G	2xD.III
1085–1094	FF33L	CHFT	Bz.III
1095–1104	FF33J	HFT	Bz.III
1105	KW Danzig	B/S	Bz.III
1106	KW Danzig	C2MG/S	Bz.III
1107–1116	Albatros W4	ED2MG	D.III
1117–1126	FF33L	C2MG	Bz.III
1127–1136	FF39	CHFT	Bz.IV
1137–1146	Brand. CC	E2MG	Bz.III
1147–1156	Travemünde F2	CHFT	D.IV
1157	Oertz W8	CHFT	Mb.IV
1158–1177	FF33L	CHFT	Bz.III
1178–1187	Brand. W12	C2MG	Bz.III
1188–1207	Rumpler 6B2	ED2MG	D.III
1208–1210	FF41A	T	2xBz.III
1211–1213	Gotha WD11	TMG	2xD.III
1214–1223	Sablat. SF5(LVG)	B	Bz.III
1224–1233	Sablatnig SF5	HFT	Bz.III
1234–1278	FF33L	CHFT	Bz.III
1279–1288	FF33L	C2MG	Bz.III

Mr. #	Type	Class	Engine
1289–1298	FF33J	HFT	Bz.III
1299–1301	LTG FD1	ED	Bz.III
1302–1326	Albatros W4	ED2MG	D.III
1327–1351	Brand. CC	E2MG	Bz.III
1352–1371	Sablatnig SF5	HFT	Bz.III
1372–1379	Gotha WD11	TMG	2xD.III
1380–1394	Brand. KDW	ED	Mb.III
1395–1414	Brand. W12	C2MG	Bz.III
1415–1430	Gotha WD14	G2MGHFT	2xBz.IVC
1431	Do. Rs.III	R3MG	3xMb.IV
1432	Staaken L	R4MG	3xD.IVa
1433	Do. Rs.II	R	4xMb.IVa
1434–1458	Rumpler 6B2	ED2MG	D.III
1459–1468	Sablat. SF5(LFG)	B	Bz.III
1469–1471	Brand. W19	C2MG	Mb.IVa
1472–1474	FF48	C2MG	Mb.IVa
1475–1477	Sablatnig SF7	C2MG	Mb.IVa
1478–1480	Brand. W	Bu	Cancelled
1481–1483	Junkers	Bu	Cancelled
1484–1513	Albatros W4	ED2MG	D.III
1514	Sablatnig SF5	BFT	Bz.III
1515–1517	Gotha WD20	G2MG	2xD.IVa
1518–1520	LTG FD1	ED	Bz.III
1521–1535	FF49C	CHFT	Bz.IV
1536–1550	FF49B	BFT	Bz.IV
1551–1553	Brand. W20	Bu	U.0
1554–1573	Brand. KDW	ED2MG	Mb.III
1574–1576	Junkers	BFT	Cancelled
1577–1596	FF33L	C2MG	Bz.III
1597–1606	FF49C	CHFT	Bz.IV
1607–1616	FF49B	BFT	Bz.IV
1617–1628	Gotha WD14	G2MGHFT	2xBz.IV
1629–1631	Gotha WD	GK2MG	Cancelled
1632–1634	Sächsische FW	C2MG	Cancelled
1635–1646	Gotha WD	GMGFT	Cancelled
1647–1649	Brand. W23	EK	D.III
1650	KW Danzig	CHFT	D.IV Cnx
1651–1662	Gotha WD14	T/G2MG	2xBz.IV
1663–1665	FF53	G2MG	2xD.IVa
1666–1668	Open		
1669–1718	FF49C	CHFT	Bz.IV
1719–1738	Albatros W4	ED2MG	D.III
1739–1741	Brand. W26	C2MGHFT	D.IVa
1742–1811	FF49C	CHFT	Bz.IV
1812–1821	FF49C	C2MGHFT	Bz.IV
1822–1823	FF59C	C2MGHFT	Bz.IV
1824–1841	FF49C	C2MGHFT	Bz.IV
1842–1856	FF49C (LFG)	CHFT	Bz.IV
1857–1871	Open		
1872–1901	FF49C(Sab)	CHFT	Bz.IV
1902–1942	Open		
1943	Gotha WD	G2MGHFT	Cancelled
1944–1945	Open		
1946–1970	Gotha WD14	G2MG	2xBz.IV

Mr. #	Type	Class	Engine
1971–1973	Travemünde F4	CHFT	Bz.IVa
1974–1979	Travemünde F2	CHFT	D.IV
1980–1999	FF33E	S	Bz.III
2000–2019	Brand. W12	C2MG	Bz.III
2020–2022	Sablatnig SF8	S	Bz.III
2023–2052	Brand. W12	C3MG	Bz.III
2053–2092	FF49C	CHFT	Bz.IVa
2093–2112	Brand. W12	C2MG	D.III
2113–2132	Brand. W12	C3MG	D.III
2133–2134	Gotha WD22	G3MG	4xD.III/D.I
2135	Travemünde F4	CHFT	Bz.IVa
2136–2137	Open		
2138	Brand. W.18	E2MG	Bz.III
2139	AEG	R	4xMb.IVa
2140–2200	Open		

Note: To simplify identification of naval aircraft, the following number groups were assigned 16 Nov. 1917:
2201–3000: Hansa-Brandenburg
3001–4000: Friedrichshafen
4001–5000: Gotha
5001–6000: Albatros
6001–6500: Sablatnig
6501–7000: LFG-Bitterfeld
7001–7500: Lübeck-Travemünde
7501–8000: Junkers-Fokker
8001–8500: Staaken
8501–9000: Zeppelin-Lindau
9001–9500: AEG

Hansa-Brandenburg Navy Aircraft Numbers 2201–3000

Marine #	Type	Class	Engine
2201–2203	W27	C3MG	Bz.IIIb
2204	W29	C3MG	Bz.III
2205	W29	C2MG	BMW.IIIa
2206	W29	C3MG	D.III
2207–2216	W19	C3MG	Mb.IVa
2217–2236	W12	C2MGHFT	D.IIIa
2237	W19	CK	Mb.IVa
2238–2257	W19	C3MG	Mb.IVa
2258	W25	ED2MG	Bz.III
2259–2278	W19	C3MG	Mb.IVa
2279–2281	Open		
2282–2284	W32	C3MG	D.IIIa
2285–2286	Brand.W-Cnx	C3MG	D.IVa
2287–2300	W29	C2MGHFT	Bz.III
2301–2302	Brand.W-Cnx	C2MGHFT	3xD.IIIa
2303	Brand.W-Cnx	R	D.I
2304	Brand.W-Cnx	R	D.II
2305	Brand.W-Cnx	G	3xD.IIIa
2306–2307	Brand.W35	G	2xMb.IVa
2308–2500	Open		
2501–2506	W29	C2MGHFT	Bz.III
2507–2536	W29	C3MG	Bz.III
2537	W19	C3MG	Mb.IVa
2538–2540	W33	C3MG	D.IVa
2541	W33	C2MGHFT	Mb.IVa
2542	W33	C3MG	BuS.IVa
2543	W33	CK	BuS.IVa
2544–2563	W19	CK	Mb.IVa
2564–2583	W29	C2MGHFT	Bz.III
2584–2587	W29	C3MG	Bz.IIIa
2588–2589	W29	C2MGHFT	Bz.IIIa
2590–2592	W-Cnx	Bu	Ur.II
2593–2642	W29	C3MG	Bz.III
2543–2652	W29	C3MG	BMW.IIIa
2653–2682	W29	C3MG	Bz.IIIa
2683–2722	W19	C3MG	Mb.IVa
2723–2725	W37	CHFT	Bz.IVa
2726	W33	C3MG	Mb.IVa
2727–2729	W34	C3MG	BuS.IVa
2730–2759	W29	C3MG	Bz.IIIa
2760–2789	W29	C3MG	D.IIIa
2790–2791	W-Cnx	Bu	Ur.II
2792–2800	W	C3MG	D.IVa
2801–3000	Open		
12001–12011	W	C3MG	D.IVa
12012–12041	W-Cnx	C3MG	Mb.IVa
12042–12051	W-Cnx	CK	Mb.IVa
12052–12054	W-Cnx	C3HFT	D.IVa

Friedrichshafen Navy Aircraft Numbers 3001–4000

Marine #	Type	Class	Engine
3001–3030	FF33S	S	Bz.III
3031–3060	FF33J	HFT	Bz.III
3061–3063	FF64	CHFT	D.III
3064–3083	FF49C	CHFT	Bz.IVa
3084–3143	FF49C	CHFT	Bz.IVa
3144–3193	FF33L	C2MG	D.III
3194–3195	FF67-Cnx	B/CHFT	D.IVa
3196	FF-Cnx	CB	D.IVa
3197–3199	FF63-Cnx	CHFT	Bz.IVa
3200	FF-Cnx	CB	D.IVa
3201–3230	FF-Cnx	CHFT	Bz.IVa
3231–3233	FF-Cnx	B/CHFT	D.IVa
3234–3300	Open		
3301–3303	FF60	G	4xD.III
3304–4000	Open		

Gotha Navy Aircraft Numbers 4001–5000

Marine #	Type	Class	Engine
4001–4002	WD28	CHFT	Bz.IVa
4003	WD28	CB	Bz.IVa
4004–4033	FF49C(Go)	C	Bz.IV
4034	WD28	CB	D.IVa
4035–4300	Open		
4301–4325	WD-Cnz	G	2xMb.IVa
4326–4328	WD27	G	4xD.III
4329–5000	Open		

Albatros Navy Aircraft Numbers 5001–6000

Marine #	Type	Class	Engine
5001–5003	W8	C3MG	Bz.IIIb
5001 (Ersatz)	W8 Replacement	C3MG	Bz.IIIb
5004–5006	Alb W-Cnx	CHFT	Bz.IVa
5007–6000	Open		

Sablatnig Navy Aircraft Numbers 6001–6500

Marine #	Type	Class	Engine
6001–6030	SF8	S	Bz.III
6031–6050	FF49C(Sab)	CHFT	Bz.IVa
6051–6080	FF49C(Sab)	CHFT	Bz.IVa
6081–6500	Open		

LFG-Bitterfeld Navy Aircraft Numbers 6501–7000

Marine #	Type	Class	Engine
6501–6520	FF33S(Rol)	S	D.III
6521–6540	FF49C(Rol)	CHFT	Bz.IVa
6541–6543	Roland V19	Bu	Ur.II
6544–6545	Roland	Bu	Ur.II
6546–6800	Open		
6801–6802	Roland	G	Mb.V
6803–7000	Open		

Travemünde Navy Aircraft Numbers 7001–7500

Marine #	Type	Class	Engine
7001–7030	F4	CHFT	Bz.IVa
7031–7040	FF49C(Trav)	CHFT	Bz.IVa
7041–7500	Open		

Junkers Navy Aircraft Numbers 7501–8000

Marine #	Type	Class	Engine
7501–7503	Cs.I	C3MG	Bz.IIIa
7504–7800	Open		
7801–7803		G	2xD.IVa
7804–8000	Open		

Staaken Navy Aircraft Numbers 8001–8500

Marine #	Type	Class	Engine
8001–8300	Open		
8301–8306		R	4xD.IVa
8307		R	4xD.VI
8308–8500	Open		

Dornier Navy Aircraft Numbers 8501–9000

Marine #	Type	Class	Engine
8501–8503	Cs.I	C3MG	Bz.IIIb
8504–8508		Bu	Ur.II
8509–8800	Open		
8801–8802	Rs.IV	R	4xMb.IVa
8803–8804	Rs.-Cnx	R	4xMb.VII
8805–8807	Gs.I	G	2xMb.IVa
8808	Cnx	R	4xMbIVa
8809	Cnx	R	4xMbIVa
8810–9000	Open		

AEG Navy Aircraft Numbers 9001–9500

Marine #	Type	Class	Engine
9001–9300	Open		
9301		R	4xMb.IVa
9302–9303		R	4xMb.IVa
9304–9500	Open		
9501–12000	Open		

German Naval Aircraft Classes

Class	Meaning
B	Two-seat aircraft with bombing equipment
BFT	Two-seat aircraft with bombing equipment and wireless transmitter
Bu	U-boat aircraft (*U-Boot Flugzeug*)
C	Two-seat aircraft with one flexible machine gun
C2MG	Two-seat aircraft with one flexible and one fixed machine gun
C3MG	Two-seat aircraft with one flexible and two fixed machine guns
CHFT	Two-seat aircraft with one flexible machine gun and wireless transmitter and receiver
CK	Two-seat aircraft with one flexible cannon and one or two fixed machine guns
E	Single-seat flying boat (*Einsitzerboot*)
ED	Single-seat aircraft with twin floats (*Einsitzer mit Doppelschwimmer*)
ED2MG	Single-seat aircraft with twin floats and two machine guns
FL	Aircraft for control of wire-guided attack boats (*Fernienkflugzeug*)
G	Twin-engine aircraft (*Grossflugzeug*)
HFT	Two-seat aircraft with bombing equipment and wireless transmitter and receiver.
R	Aircraft with three or more engines (*Riesenflugzeug*)
S	Two-seat training aircraft with dual controls (*Schulflugzeug*)
T	Torpedo aircraft (*Torpedoflugzeug*)
U	Practice aircraft (*Uebungsflugzeug für Alleinflieger*)
V	Test aircraft (*Versuchsflugzeug zur Erprobung von Motoren, Propellern, Instrumente, usw.*)

AEG 1914 Flying Boat

Left: The AEG monoplane flying boat completed in July 1914. Only one was built; the Navy evaluated the aircraft and judged that it was unsuitable for naval service. The rod mounted above the engine was for lifting the boat out of the water. Power was by a 150 hp Benz Bz.III engine. Although a monoplane, there are distinctive strut braces under the wing.

AEG S1

Above: The unsuccessful AEG S1 floatplane being prepared for flight tests in May 1914 on the Havel See near the Hennigsdorf factory. The wings could be folded back for storage. Despite being powered by a 150 hp Benz Bz.III engine, the S1 was too heavy and had poor climb; it was not accepted by the Navy. It has 2½ bays of bracing.

On 12 October 1917, the German Navy ordered a giant AEG R-flying boat (Marine Number 9301, formerly Marine Number 2139) powered by four 245 hp Maybach Mb.IVa engines. The AEG flying boat project was listed as 'in design' until the project was canceled on November 26, 1918.

Two AEG R-floatplanes (Marine Numbers 9302 and 9303), powered by four 245 hp Maybach Mb.IVa engines, were ordered by the German Navy on January 23, 1918. The giant seaplane was basically an AEG R.I fitted with floats. The AEG 9302 fuselage, control surfaces, and metal fixtures were under construction in June. At the time of cancellation on December 18, 1918, the AEG 9302 fuselage was in final assembly and the floats and beaching trolley, both manufactured by the Zeppelinwerk Staaken, were completed. Only the fuselage parts and metal fittings for AEG 9303 were finished at the time.

AGO Pusher Seaplane Trainer

Above: The Ago Seaplane Pusher Trainer was a seaplane derived from the Ago Pusher trainer. Floats were added under the fuselage nacelle and under the tailplane and the wings are extended to 4½ bays of bracing. The additional wing area was needed due to the additional weight, especially during a water take-off. (Peter M. Grosz Collection/STDB)

Below: Ago Floatplane Pusher Trainer Marine Number D15 photographed February 2, 1914 while being refueled by hand. The wheels have been removed. The twin tail floats are distinctive. (Peter M. Grosz Collection/STDB)

AEG Biplane Seaplane

Above: The Ago Seaplane Biplane (number 9) with another in the background. Powered by a 150 hp Argus As.III, the few purchased by the Navy were assigned Marine Numbers 65–69. It has twin fins and 3½ bays of bracing. (Peter M. Grosz Collection/STDB)

Ago Tractor Biplane Floatplane Trainer

Below: The Ago Tractor Biplane Floatplane Trainer wasa 3-bay design built in 1913 and powered by a 100 hp Gnôme rotary. A few purchased by the Navy were assigned Marine Numbers 30–39. (Peter M. Grosz Collection/STDB)

Above & Below: The Ago Tractor Biplane Floatplane Trainer was built in 1913; here it is in the Ago factory. Note the engine cowling and tailplane differences between this example and the one on the facing page. Such detail changes between different aircraft of the same type were common on early aircraft.

AGO C.Iw

Above: Ago C.Iw (MN 115) on a railway flatcar in front of Hangar 3 on the mole at Zeebrugge. The tail booms are made of a light birchwood veneer and only appear dark in the photo due to the film used. There are 3 bays of bracing. (M. Szigeti)

Above: Ago C.Iw (MN 115) was a C.I with floats instead of wheels attached via a robust assembly of struts. One aircraft, MN 115, was built and eventually assigned to Zeebrugge. Power was a 150 hp Benz Bz.III.The Ago C.Iw was one of the few German pusher designs. (Peter M. Grosz Collection/STDB)

AGO C.IIw

Above & Below: Ago C.IIw MN 539 was the first of two aircraft built; the second and last was MN 586. The C.IIw was a floatplane development of the 3-bay AGO C.II. The engine was a 200 hp Benz Bz.IV. The enlarged wing and greater engine power were intended to improve performance and payload. Several men in the original photograph above have been crudely erased. (Peter M. Grosz Collection/STDB)

Albatros MZ2 Floatplane

Above: The Albatros MZ2 landplane was also built as a floatplane; there a float-equipped MZ2 is being prepared to be lowered into the water by crane. This version had twin floats equipped with wheels as an amphibian. (Peter M. Bowers Collection/The Museum of Flight)

Above: This Albatros MZ2 floatplane has a central float for buoyancy and primitive wing-tip floats for stability. (Peter M. Bowers Collection/The Museum of Flight)

Albatros *Wahl*

Above & Below: The Albatros *Wahl* (Whale) flying boat featured a hull built by Naglo Werft. It was powered by a 100 hp Mercedes in the hull that drove a pusher propeller via a drive shaft. From the rear it resembled Oertz designs. It was not successful and only a single prototype was built. (Peter M. Bowers Collection/The Museum of Flight)

Albatros *Taube* Floatplanes

This Page & Facing Page: Like most German pre-war manufacturers, Albatros experimented with the monoplane *Taube* configuration. These photos show the HE Amphibian – Bodensee, the Albatros *Taube* floatplane designed for the Bodensee competition that was won by Hirth in this aircraft. While the *Taube* configuration had no future, the fuselage structure Albatros developed for this aircraft strongly influenced the early Albatros fighters and was a key to their success. (Peter M. Bowers Collection/The Museum of Flight)

Concorso Internazionale Idrovolanti - Como - Ottobre 1913
L'Albatros caduto a Mornello
Albatros 6

Albatros Float Biplane *Bodensee*

Above & Below: A straight-forward floatplane development of the new Albatros DE biplane resulted in the *Wasser-Doppeldecker* Bodensee for pre-war contests. (Peter M. Bowers Collection/The Museum of Flight)

Albatros WDD

Above & Below: Originally known as the Albatros WDD (for *Wasser Doppeldecker Doppelsitzer,* water biplane two-seater), the WDD was a floatplane derivative of the DD land plane. The WDD was retroactively designated Albatros W.1. The WDD, which had 3 bays of bracing, first flew in the autumn of 1913. The example below has a different rudder shape.

Above & Below: Side view of the Albatros WDD flown in the 1913 Bodensee Contest shows the rudder shape more clearly. (Peter M. Grosz Collection/SDTB)

Albatros W.1

Above: Albatros W.1, Marine Number 552, identified as a Brandenburg built by Danzig in the list of marine numbers. The rudder and fin are modified, the tail float has been deleted, and side radiators are fitted.

Below: Closeup of an Albatros W.1. (Peter M. Grosz Collection/SDTB)

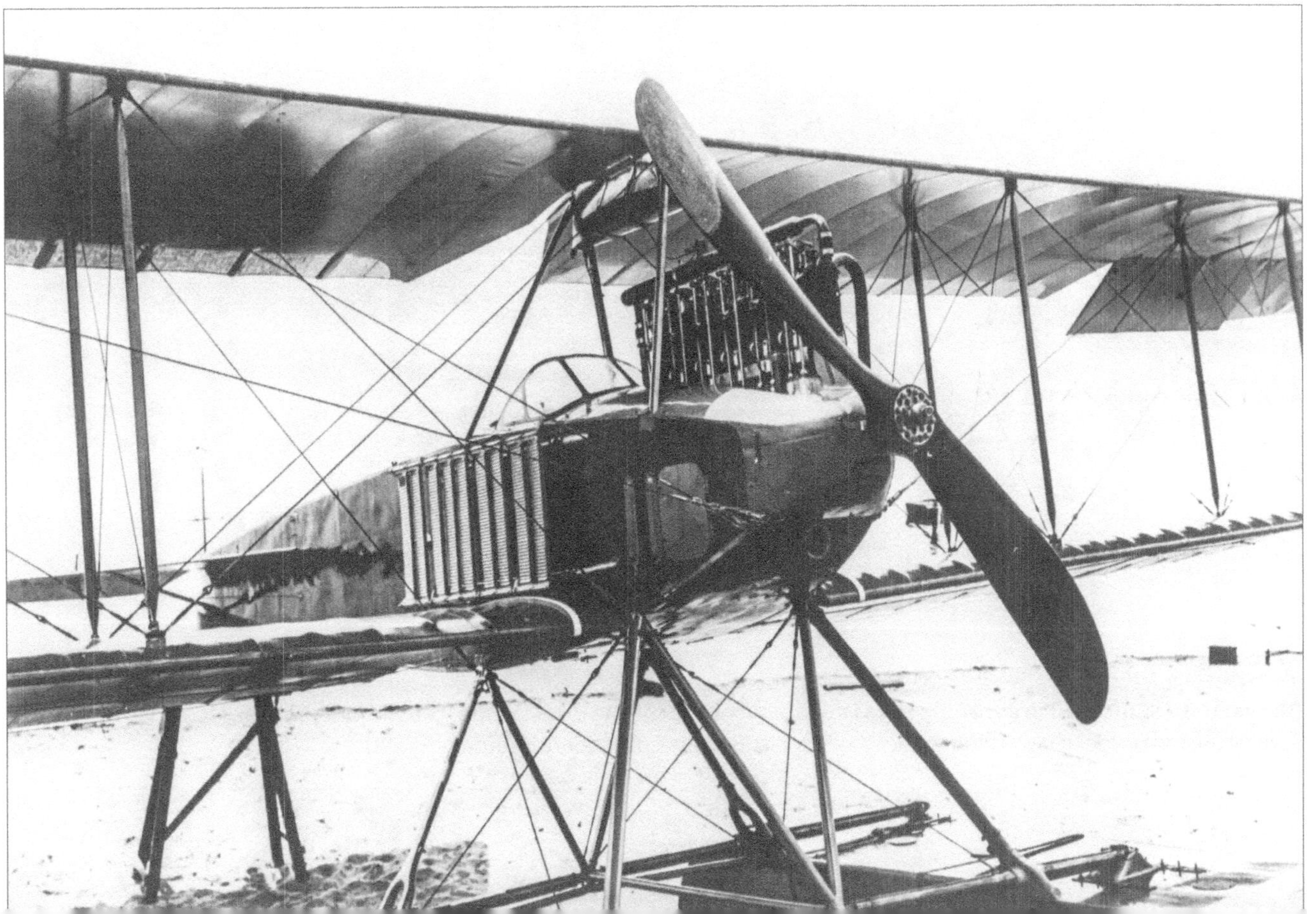

Above: This Albatros W.1 is seen just after take-off; it has no tail float but does have a leading edge radiator and carries national insignia and a naval pennant. (Peter M. Grosz Collection/SDTB)

Above: This Albatros W.1 in its hangar has a later design of enlarged vertical tail surfaces without tail float. Note the variety of vertical fin and rudder shapes fited to the W.1 (Peter M. Bowers Collection/Museum of Flight)

Albatros W.2

Above & Below: The Albatros W.2 was a clean, 2-bay design for a two-seat reconnaissance floatplane. It was derived from the Albatros C.III airframe and was powered by a 160 hp Mercedes D.III; streamlined side radiators were fitted. The observer had a single flexible machine gun, but there was no fixed gun for the pilot.

Albatros W.3

Above: Marine Number 527 was the only Albatros W.3 built. The W.3 was a class TMG, a torpedo bomber armed with two flexible machine guns. The wood construction was the same technology applied to the Albatros D.I.

Below: The completed Albatros W.3 was a clean design with minimal frontal area for a fairly large, twin-engine biplane floatplane. Albatros had learned the importance of streamlining for performance and applied that lesson to the W.3. It had 2 bays of bracing outboard of the pusher engine nacelles for its 150 hp Benz Bz.III engines. The radiators were mounted high in the front of the nacelles. (Peter M. Grosz Collection, SDTB)

Albatros W.4

Above: Albatros W.4 MN 964 of the second production batch. The W.4 was a floatplane fighter developed from the D.I fighter. The wings were enlarged to provide more lift to carry the floats. The prototypes and first production batch had a single machine gun and ear radiators. The second batch and later had two guns, and the 5th batch an later had airfoil radiators and ailerons on all wings. (Peter M. Grosz Collection/SDTB)

Below: With airfoil radiator, four ailerons, and two guns, MN 1486 is in the final W.4 production configuration. (Peter M. Bowers Collection/Museum of Flight)

Albatros W.5

Above: This view of Marine Number 845 also shows that ailerons were fitted to the upper wing only on this aircraft. It is not known if this aircraft was later upgraded to the standard four-aileron configuration shown in the SVK drawing. The W.5 torpedo bomber was a refined development of the W.3 with the same span and length and powered by the same 150 hp Benz Bz.III engines, but with enlarged floats. The car-type radiators were improved for lower drag with better cooling. Initially the W.5 had ailerons on the upper wing only like the W.3, but for better controllability the final control configuration was ailerons on all wings with actuating rods between upper and lower ailerons. A batch of five W.5 torpedo bombers, MNs 845–849, were ordered the same month as the W.3 was delivered.

Above: This view of W.5 Marine Number 847 with torpedo fins still visible below the fuselage. The rudder has a straight trailing edge and is more angular than that on MN 845. Ailerons were fitted to all wings with actuating struts connecting them. (Peter M. Bowers Collection/Museum of Flight)

Albatros W.8

Above: The first prototype Albatros W8 was a handsome, well-streamlined two-seat fighter powered by a V-8 that was not yet in mass production. However, the flat radiator under the wing somewhat spoiled the overall streamlining. The ailerons had horn balances, unusual for an Albatros design.

Below: MN 5002, the second Albatros W8 prototype, shows that the elegant spinner of the prototype was abandoned but the drag-producing block radiator beneath the upper wing was retained. The aerodynamic horn balanced ailerons were replaced with plain ailerons. (Peter M. Bowers Collection/Museum of Flight)

Brandenburg W

Above: The Type W was a crude looking machine but reflected the state of the art at the time. The ungainly pontoons and 3- bay bracing are noteworthy. MNs 57–58, 71–72, 231–235, 260–263, 422–423, and 484 were powered by the 150 hp Benz Bz.III. MN 116 had a 140 hp Argus, and MNs 264–273 had the 150 hp Maybach Mb.III.

Above: Brandenburg Type W.

Brandenburg NW

Above: A fine study of an NW biplane in flight. This early model has a tail float. The 160 hp Mercedes was the most common engine, although K148 used by Austria had a 165 hp Rapp.

Above: NW 517 had a plain rudder with no horn balance. There were 3 bays of struts.

Below: NW 602 from the fourth batch of NW floatplanes ordered. The rudder has a horn balance but there is no tail float.

Brandenburg LW

Above & Below: LW Marine Number 571. Note the machine gun mount in the rear cockpit and the slinging cable on the top wing center section. Powered by the 160 hp Mercedes D.III engine, three aircraft were built; MN 477 (category CBMG), 485 (category B), and 571 (category C), and were delivered in August 1916. The interplane struts for the 2-bay bracing were sloped inward at the top and the rudder had a typical Brandenburg horn balance.

Brandenburg GNW

Above & Below: Brandenburg GNW MN 625. The GNW was an unarmed, 3-bay development of the NW reconnaissance seaplane. The prototype had a 160 hp Mercedes D.III but production aircraft had the 150 hp Benz. The rudder had the typical Brandenburg horn balance.

Below: Brandenburg GNW MN 627 survived into 1918 to receive the straight sided crosses.

Brandenburg FB

Above: Brandenburg FB MN 511 was the prototype. The rudder is small and lacks the typical Brandenburg horn balance. The MN has been remarked in the usual size and style on the prototype FB seen here at the Gotha works. The prototype had a 165 hp Austro-Daimler engine. Germany had six examples and the Austrians used many more. Later examples used the 175 hp Rapp or 160 hp Mercedes D.III. The FB had two or three crew with a gunner in the front cockpit with flexible machine gun. The interplane struts were in a V-shape.

Above: Brandenburg FB MN 514 in water with Friedrichshafen FF33 MN452 in background. This FB has a horn-balanced rudder and MN 511 was refitted with a similar rudder.

Brandenburg GW

Above: The prototype Brandenburg GW torpedo bomber was allocated Marine Number 528. The two bays of interplane struts were sloped inward at the top like many Brandenburg designs. The twin engines were 160 hp Mercedes D.III. The prototype had triple fins and rudders.

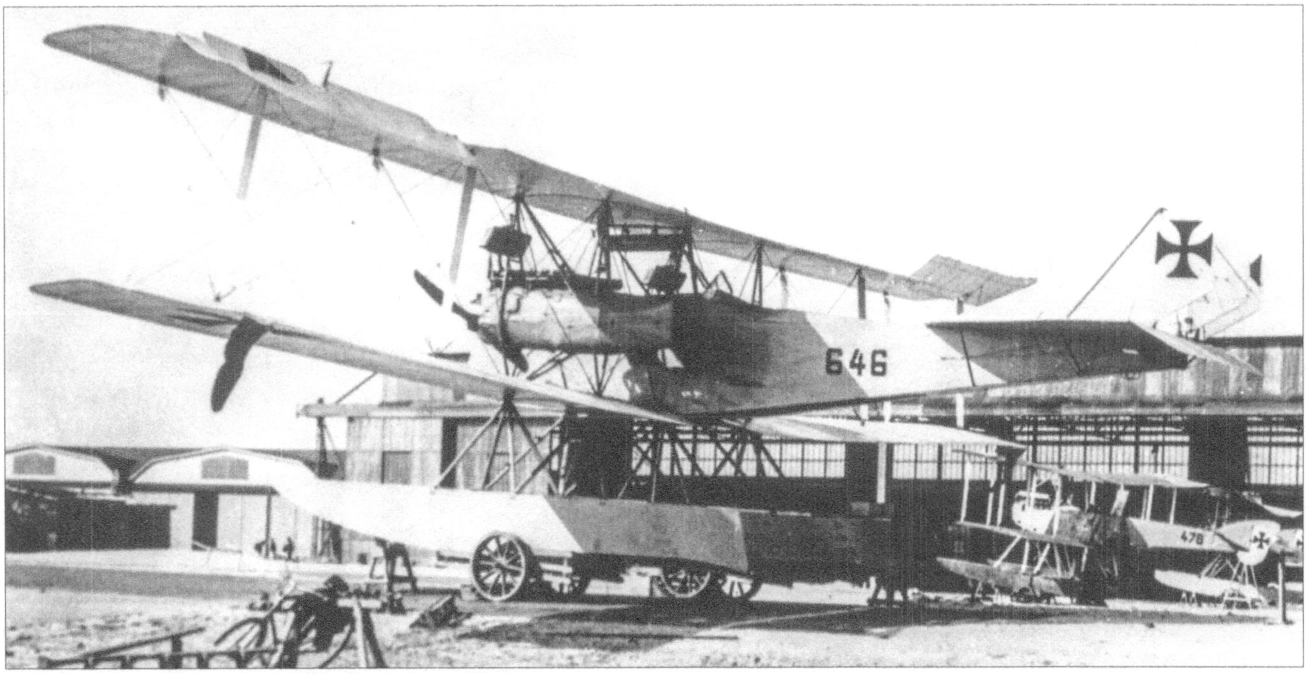

Above: MN 646, the first GW of the second production batch. Gotha WD8 MN 476 is in the background at right. MN 646 had the triple rudders of the first production batch, but MN 650, the last aircraft of the second batch, had the single rudder and aileron servo tabs used on the later production batches, although it retained the forward gun turret. Moving the float dollies of these big aircraft required a lot of manpower.

Above & Below: The more rounded nose with the gunner's position moved to the rear and single large rudder of the third production batch are well displayed in these photographs of Brandenburg GW MN 700. Also significant was the addition of aileron servo tabs on this batch, which reduced the pilot's control forces. A torpedo was loaded.

Right: Brandenburg GW MN 650 with torpedo. Given the location of the pilot it is understandable that the gunner's position was relocated to the nose to improve his view.

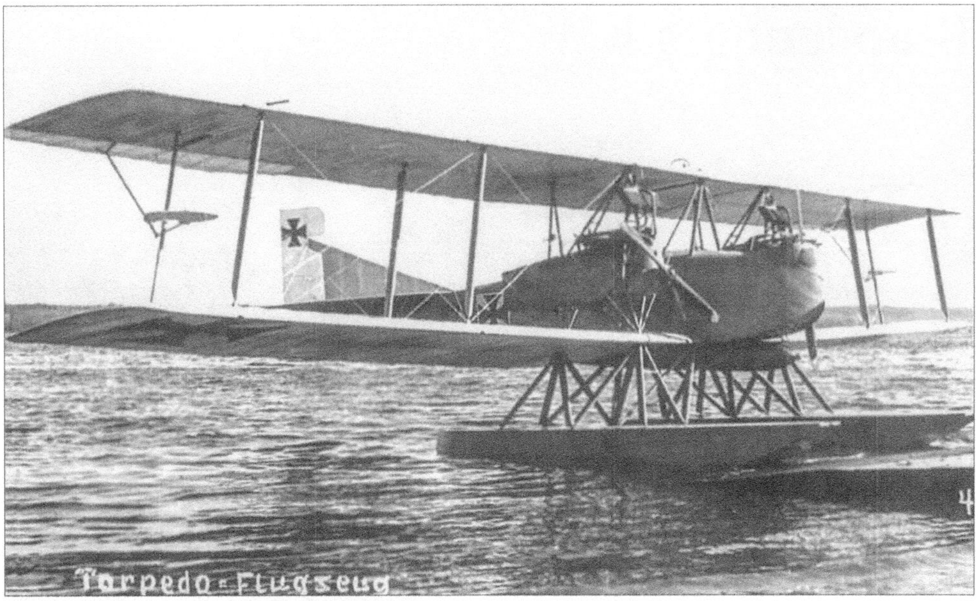

Left: A Brandenburg GW from the fourth production batch (MN 1080–1084) showing the single rudder, balanced elevators, and shorter fuselage. This version could carry a 1,600 lb torpedo. There appears to be a pitot tube suspended from the nose of the fuselage. The single rudder had horn balances top and bottom.

Right: Brandenburg GW MN 701 wears an unusual and distinctive camouflage and markings scheme. This was the GW purchased by the *kuk Kriegsmarine* and appears to be at one of their naval air stations; a flying boat with red/white/red Austro-Hungarian markings on its wing is in the right background.

Brandenburg KW

Above & Bottom: Brandenburg KW Marine Number 589 fitted with a four-bladed airscrew for its 200 hp Benz Bz.IV. Three KW two-seat, three-bay reconnaissance biplanes in the CFT Class were supplied during April and May 1916. The KW featured the inward-sloping interplane struts and a first bay devoid of bracing cables. Although intended to have the 260-hp Mercedes D.IVa engine after testing, the 200-hp Benz Bz.IV replaced the Mercedes and the performance suffered accordingly, such that the machines, MNs 588–590, were only used for training.

Left: Brandenburg KW MN 590 in full camouflage and markings. This was the machine used for the SVK "Atlas" three-view and these photographs are thought to have been taken at Warnemünde.

Brandenburg GDW

Above: The Germans tried to develop a practical torpedo floatplane and the GDW was one of the very large floatplanes built in pursuit of this goal. Power was from two 200 hp Benz Bz.IV engines.

Right & Below: The GDW was one of the very large floatplanes built to develop a practical torpedo floatplane; only one, MN 746, was constructed.

Brandenburg KDW

Above: This striking view of Marine #748, the first prototype, shows the washed out ailerons. Many WWI airplanes had the angle of attack of their ailerons reduced or 'washed out' to ensure the inner part of the wing stalled before the outer part, where the ailerons were fitted. With no 'wash out', the outer part of the wing would stall first and the pilot would lose aileron control at the beginning of the stall, making a spin much more likely. The metal fairings enclosing the apex of the wing star-struts of #748 are not seen on later aircraft. The prototypes had 150 hp Benz Bz.III engines and frontal radiators.

Above: Marine Number 783, the second KDW prototype, showing its frontal radiator. The machine gun is far forward, out of the pilot's reach in case of jams. The wing structure is evident through the fabric. The *Balkenkreuz* on the wings appear to be painted over white backgrounds. Although it is difficult to see in this view, there is no fixed fin above the fuselage.

Above: The first KDW from the initial production batch, Marine #912 went to Zeebrugge. On 10 May, 1917, this aircraft was wrecked while being flown by *Oblt.z.S* Kurt Reinert, who later died from the injuries he sustained. *Lt.* Hammer downed Sikorsky Mourometz *IM-6* with MN 748; he complained about the forward location of the machine gun and it is now located closer to the cockpit. An airfoil radiator has been fitted to the upper wing, just right of center.

Above: Another KDW from the first production batch was Marine #921, which initially went to Libau, arriving there on 24 Feb., 1917. This aircraft was apparently at Putzig by May, and was wrecked while being flown by *Lt.* Markwald on 11 August, 1917. This photo was taken in front of a Hansa-Brandenburg hangar, apparently prior to June, 1917, when additional bracing struts were ordered to be installed on all KDWs. 150 hp Benz Bz.III engines powered the first batch.

Above & below: These photos, Marine Number unknown, clearly shows the light interplane struts ordered to be added to all KDWs in June, 1917. The light struts run from the front spar of the lower wing to both spars of the upper wing and increased the torsional stiffness of the upper wing, which improved aileron response. Compared to the prototypes, the machine gun has been moved back closer to the cockpit to give the pilot better access in case of jams. A fixed vertical fin has been added above the fuselage, which was not done on early production batches. An airfoil radiator was fitted. Later production aircraft had 160 hp Mercedes, 160 hp Maybach, or 150 hp Benz engines.

Above: Marine Number 1562 of the final KDW production batch displays the additional interplane struts and shows off the three-color hexagonal camouflage fabric used on the upper surfaces of German naval aircraft late in the war. The machine guns have been moved up in front of the pilot for better access. This final production batch had 160 hp Maybach Mb.III engines.

Above: A KDW from the last production batch, #1562 was one of the aircraft originally sent to *II.Seefliegerabteilung*, presumably for assignment to a western front unit. Following a series of crashes at the *Seekampfeinsitzerschule* at Putzig, this and several other *II.SFA* machines were ordered to be transferred to *I.SFA* and sent to the Putzig school. Standard late naval lozenge camouflage was applied to this machine. The fixed vertical fin above the fuselage is enlarged compared to the prototypes and early production machines.

Brandenburg CC

Above & Right: The prototype Brandenburg CC after application of the *Marine Nummer* 946 and national markings with single-gun armament fitted. It is often hard to distinguish the MN against the dark varnished hulls. In this case it is painted at the rear of the hull in front of the cross. Lower wing crosses do not have any outline. The cross panels on the upper wing can be discerned through the fabric as a darker tone. Star-struts were used and no engine cowling was fitted.It had a 150 hp Benz Bz.III.

Left: On MN 1144 the number is painted on the nose under the cockpit. 1144 is from the first production batch allocated to the German *Kriegsmarine* and has twin guns and a cowled 150 hp Benz Bz.III.

This Page: Brandenburg was not immune to the triplane craze. Note that the serial is in the form A.45., indicating that it was built by Brandenburg rather than a modification carried out by the *kuk Kriegsmarine*. However, the *Typenshchau*, the book produced by Heinkel's own company, lists the machine as the Phönix Dr.I. That is because the modification was most likely to have been made to an existing standard CC airframe at Phönix Werke in Vienna and not by Hansa-Brandenburg in Germany. Austrian CCs had 160 hp or 185 hp Austro-Daimler engines.

Above: MN 1348 was from the second batch produced for the German Navy. It has the extra interplane struts that were to be applied to all star-strutter types. These struts were to improve the aileron effectiveness by strengthening the upper wing in torsion. Note the forward sight on the decking in front of the cockpit and the uncowled engine.

Brandenburg W.11

Above: If not for the Marine Number 988 the W.11 can be mistaken for a KDW fighter retaining the "star strutter" bracing of its predecessor. Larger than the KDW, twin guns and a 200 hp Benz Bz.IV were fitted.

Above: Rear quarter view of W.11; from this vantage point it still looks like a KDW. Three aircraft, MN 988–990, were built; all were assigned to Zeebrugge from March 1917.

Above: The W.11 was a slightly enlarged, more powerful version of the KDW fighter retaining the "star strutter" bracing of its predecessor and looks virtually identical; the Marine Number is the fool-proof way to tell the difference.

Brandenburg W.12

Above: The prototype W.12 floatplane prototype MN 1014 in January 1917, outside the Brandenburg factory hangar at Briest. The prototypes differed markedly from the production examples. The pilot's cockpit has a low side to starboard to facilitate entry, and the gun ring is small without the bulged sides for the standard gun ring. Note the light color of the bottom of the pontoons. The synchronised machine gun fitted to the W.12 was the Maxim IMG-08 with its distinctive fat jacket. Later machines carried two guns but the extra weight had a detrimental effect on performance.

Below: W.12 MN 1012 on the Zeebrugge Mole; the single-bay bracing and leading-edge radiator are distinctive.

Above: W.12 MN 2008 with crew after a successful forced landing on land! The personal insignia is a lucky four leaf clover and it worked in this situation. This aircraft has a nose radiator. The strong float bracing is evident. The fuselage and rudder cross have a white outline. A distinguishing feature of these Brandenburg two-seat floatplane fighters was the underslung horn balance on the rudder with the lack of fixed vertical surfaces compensated for by the deep fuselage. This design feature was to give the gunner the best field of fire. The elevators also feature horn balances. There were three W.12 prototypes and seven production batches; depending on the batch these were built with 150 hp Benz Bz.III, 160 hp Mercedes D.III, and 170 hp Mercedes D.IIIa engines with both leading-edge and nose radiators. The W.12 was category C2MG except for MN 2023–2052 that were C3MG and MN 2217–2132 that were C2MGHFT.

Above: MN 1401 in flight above a sea of clouds.

Above: Famous photograph of MN 1414 in flight showing the hexagonal camouflage applied to upper surfaces. Pre-printed fabric was used; however, it is thought that early applications of this scheme may have been hand painted, but confirmation is lacking. The sides and presumably the bottoms of the floats are black bituminous paint. 1414 was originally going to be shipped to Turkey but instead was sent to Zeebrugge. The German Navy's specifications stated that the wing crosses had to be positioned at the tips of the wings; however, Brandenburg painted them in the same locations as it would for aircraft delivered to the Austro-Hungarians. 1414 was severely damaged in a bombing raid by No.211 Squadron, RAF, during the summer of 1918, that damaged other machines from 3C *Staffel*.

Below: The first van Berkel WA. It was a Dutch copy of the W.12.

Above & Below: MN 2016 with a 195-hp Benz Bz.IIIb V-8 engine, 14 March 1918. The V-8 engine was installed for testing. This machine has all the modifications to date including four ailerons, large cut-out in the upper wing center, and longer fuselage. Points to note are the stencil on the nose of the float; the dark color of the struts; the fuel leads from the upper wing gravity tank, and the application of the national insignia with white outline. MN 2112 appears in the background below. In the hangars in the background above are a set of huge wings. MN 2016 served at the SVK till the war's end but was never formally accepted by the Navy.

Brandenburg W.13

Above: The Brandenburg W.13 was designed in Germany for construction in Austria. The engine was a powerful 350 hp Austro-Daimler. The crew of three included a gunner with flexible machine gun in the front cockpit. The robust interplane struts were in a V-shape. K.307 on the turntable at Brindisi displays the period after the "K" indicating the components were built in Briest, Germany. The red and white wing tip and tail markings are well shown. K307 was captured and the subject of a British report on the type.

Above: Five 50-kg Skoda bombs are mounted on the hull of K373. Accepted for service on 12 September 1917, at Trieste.

Brandenburg W.16

Above: The Brandenburg W16 was an attempt to produce a replacement for the KDW by improving its flight characteristics. Unusually, a 160 hp Oberursel U.III twin-row rotary was fitted. The W.16 featured a different strut arrangement than the KDW to achieve better field of view for the pilot. Three aircraft, MN 1077–1079, were built, with MN 1078 shown. Appearance of the highly effective W.12 two-seat floatplane fighter eliminated any possibility the W.16 would be produced in quantity.

Above: The W16 featured a different strut arrangement than the KDW to achieve better field of view for the pilot.

Brandenburg W.17

Above: Testing of the W.17 by Brandenburg personnel before shipment to Austria-Hungary. The development from the Type CC flying boat with its star-struts is evident in this view. The lower wing was cantilevered, resulting in the long, sloping struts from the hull to the upper wing.

Above: The W.17 under test by Brandenburg personnel before shipment to Austria-Hungary. Its development from the Type CC flying boat is evident in this view. Gottfried Banfield, the leading Austro-Hungarian naval ace, judged that the cantilevered lower wing was not sturdy enough for maritime operations.

Brandenburg W.18

Above: The German Navy W.18. The machine has the late-war hexagonal camouflage painted on the hull turtledeck. The Marine Number appears to be marked crudely on the forward hull as 2138. Note the location of the twin machine guns. The W.18 was designed to replace the CC after failure of the W.17; the main visual and structural difference was the W.18 featured a conventional single-bay of bracing. The engine was also more powerful; different batches had either a 200 hp Hiero or a 230 hp Hiero. MN 2138 was the only W.18 in the German Navy; the Austro-Hungarian *k.u.k. Kriegsmarine* bought 47 examples.

Above: This W.18 is in the hexagonal camouflage to wings and hull. It has a small unbalanced rudder. It may be MN 2138 but late production boats for Austria-Hungry were in this color scheme.

Brandenburg W.19

Above: This unmarked W.19 was certainly the first prototype. A key indication this was the first prototype W.19 is the lack of ailerons on the lower wing; all other W.19s had ailerons on both upper and lower wings for improved maneuverability. The horn exhaust extends horizontally from the engine cowling.

Above: Marine Number 1470 was the second prototype W.19. The aircraft is painted and marked and the engine cowling appears to be designed for individual exhaust pipes. The float bracing appears very sturdy. Only the leading edge radiator spoils its clean lines.

Above: The W.19 was an enlarged version of the W.12 designed for greater range and endurance to similar configuration. The most visible difference between the types is the W.19 was a 2-bay design while the W.12 was a single-bay design. The W.19 had a streamlined nose and spinner for its 240 hp Maybach Mb.IVa engine, another point of difference between the types. Although the prototype had wire bracing for all bays, production W.19 aircraft had no wire bracing in the inner bay of bracing to give an unobstructed field of fire forward for the gunner. This view of Marine Number 2207 emphasizes its streamlined fuselage and the robust float bracing. The fairing over the pilot's left gun shows dedication to streamlining and protecting the gun mechanism from the airflow and especially salt-water spray. The W.19s were finished in the standard late-war naval camouflage specified by the Navy; the three-color naval lozenge fabric covering the upper surfaces of the wings and fuselage are just visible. This camouflage fabric was also used on the top of the tailplane.

Above: This side view of Marine Number 2207 emphasizes the gunner's clear field of fire. The actuating strut between the ailerons on upper and lower wings has now been streamlined to reduce drag.

Brandenburg W.20

Left: Marine Number 1552 was the subject of the SVK test and drawings. This machine has conventional wing bracing and represented the final configuration for the type. The W.20 was an unarmed, single-seat reconnaissance flying boat for use from submarines.

The W.20 was the first of several designs for small flying boats that could be disassembled for storage on a submarine and quickly erected for use. The W.21 and W.22 were subsequent designs in this series.

Right Above: The First W.20, MN1551, with cantilevered lower wing separate from the upper wing supports.

Right Below: This is thought to be 1551 with the upper wing of less span than the lower. It appears to have a large amount of dihedral to the lower wing. Note the wing support struts conform to those shown on the Brandenburg three-view drawing. Alternatively it may be the full size mock-up built to test the feasibility of the proposal.

Brandenburg W.21 & W.22

Above: Identified as the W.22, this flying boat is definitely related to the CC series. The wings have the late bracing arrangement with the extra support struts and the wing surfaces appear to be the same as the CC. The hull has been modified with sponsons. It is thought that this was an experimental machine to test the new hull but no data are available.

Brandenburg W.25

Above: The final development of the KDW configuration was the W.25. It has three-color naval hexagonal fabric on the upper surfaces; conventional interplane struts replaced the 'star-strut' arrangement of the KDW.

Above & Right: The W.25 had the fixed upper vertical fin of the late-production KDW, but differed in having ailerons on all four wings, which also featured increased span and area. These changes likely improved its handling characteristics, but not enough for a production order.

Below: This view of the W.25 shows the careful attention to minimum frontal area and streamlined nose entry to minimize drag.

Brandenburg W.26

Above: The early crosses on this W.26 indicates that it is the first machine, MN1739, of the order for three.

Above & Below: These photographs show a W.26 in full marine camouflage with the MN marked on the fuselage forward of the cross. The MN can be made out on the original print on the interplane struts as 1739. No spinner was fitted.

Brandenburg W.27

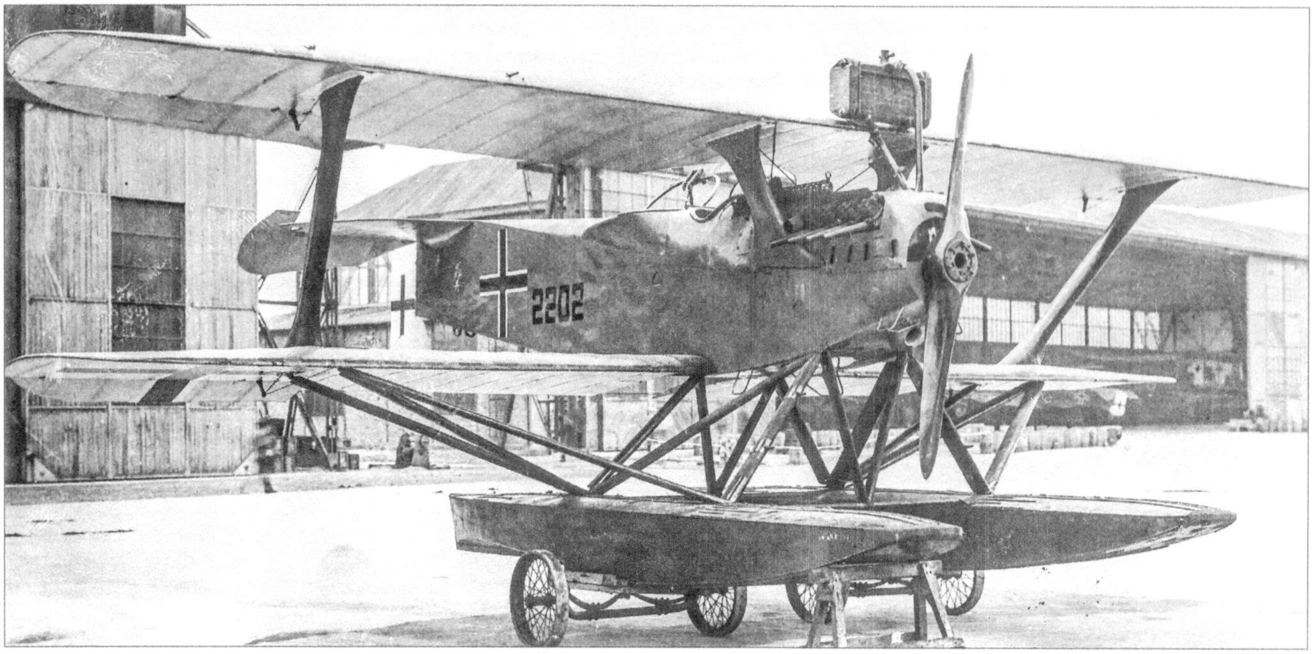

This Page: Only photographs of W.27 MN 2202 have survived. Brandenburg plans show a neatly cowled engine with spinner. The W.27 was a development of the W.12 powered by an experimental 195 hp Benz Bz.IIIb V-8.

Brandenburg W.29

Above: The prototype W.29 on the Planer See, 27 March 1918. The small early rudder was to be significantly increased in area in the prototype and later production aircraft.

Above: The prototype W.29 with its Marine Nummer 2204 applied, in the configuration in which it was accepted for service with enlarged rudder and was subject of the *SVK* drawing for the type. The machine is in the standard color scheme of pre-printed fabric on the upper surface of the wings and tailplane. Fuselage and sides and bottom painted light blue-grey. The turtledeck behind the gunner's cockpit has been painted to match the wing fabric or else pre-printed fabric has been doped to the ply. The curved crosses are marked with a white outline to the fuselage, rudder and probably the top surface of the wings. The underwing crosses are applied to white panels. These photographs were dated 4 April 1918, at Warnemünde. The ailerons are now fitted with balance surfaces.

The W.29 was a monoplane development of the W.12 biplane with the purpose of increasing its speed. It was successful and was produced in substantial numbers. Various engines were used for different production batches, inluding the 150 hp Benz Bz.III, 160 hp Mercedes D.III, 170 hp Mercedes D.IIIa, and 185 hp BMW.IIIa.

Above: This W.29 appears to be 2512 from what can be discerned of the Marine Nummer. Christensen is in the cockpit and it bears his personal insignia and early production tailplane. The observer has a Parabellum 14/17 light machine gun with an (Oigee) telescopic sight mounted in the production style cockpit with gun ring.

Below: Austrian C.I photographed at Csepel Island in October 1918. The radiator was typical of those used by the Austro-Hungarians on the 185-hp Austro-Daimler engine. It was a W.29 built under license.

Brandenburg W.32

Above & Below: W.32 MN 2282, presumably at Warnemünde for SVK testing of the type. The MN is repeated in small numerals on the bottom of the rudder, and on other components. The W.32 was a handsome machine with its neatly cowled engine and spinner. Powered by a 160 hp Mercedes D.III engine, it was developed from the W.27.

Brandenburg W.33

Above: Despite being a larger aircraft than the W.29, the W.33 is almost indistinguishable from the W.29 in photographs and if the lack of a cut-out at the wing root cannot be seen, then it is necessary to identify the Marine Nummer to be sure that the particular floatplane is indeed a W.33. MN 2538 was ordered as one of the three prototypes on 24 April 1918. This particular machine was in the CK category and was equipped with two fixed, synchronised machine guns for the pilot and a 2 cm Becker cannon for the observer. The engine was the more powerful 260-hp Mercedes D.IVa.

Above & Below: More views of W.33 MN 2538 taken at the *SVK* test center at Warnemünde on 30 August 1918, emphasize its clean lines for such a large floatplane.

Above: The W.33 had a successful postwar career in Norway. F.20 II in the dark green finish. This machine has two Lamblin radiators under the fuselage. The metal badge of the Marinens Flyvebaatfabrik can be seen on the fuselage near the footstep under the pilot's cockpit. Delivered on 27 August 1923, this seaplane was used together with F.28 to search for herring along the west coast in November 1929. The operation lasted until the end of December and in January the tow aircraft returned to Horten. In 1930 an enclosed cabin for a radio operator was fitted. F.20 II was written off after a fatal crash in August 1933, the crash leading to the grounding of the Brandenburgs still in service.

Above: This ski equipped A.22, (license-built W.33) thought to be 4.E.24 of the Finnish Aviation School circa 1925, has only a single underslung Lamblin radiator. This machine was in service by 8 December 1923 and survived until 3 July 1925, when it was wrecked. Total flight time 154:50 hours.

Brandenburg W.34

The W.34 was a slightly larger version of the W.33. Three were ordered, MNs 2727–2729, but only MN 2727 was completed, the other two were waiting their 300-hp Basse und Selve BuS.IVa engines when the war ended. MN 2727 was taken over and tested by the French post-war but no documents as to the trials has been found to date. According to Green and Swanborough some were fitted with 300-hp Fiat A.12bis engines after the Armistice but this has not been confirmed. Even Heinkel's *Typenschau* lists the Fiat engine. The post-war versions of the W.33 built in Norway and Finland used the Fiat engine and this may have led to the statement that it was used in the W.34.

Brandenburg W.35

The W.35 was a large flying boat, designed around two 300-hp Basse und Selve engines, to counter the large Curtiss and Felixstowe flying boats operated by the British in the North Sea. The W.35 was a large monoplane designed for a crew of four, including pilots side by side, and its armament was to include a 2cm Becker cannon and bombs. Two hulls were almost complete by the end of October 1918, but the boats were destroyed after the Armistice.

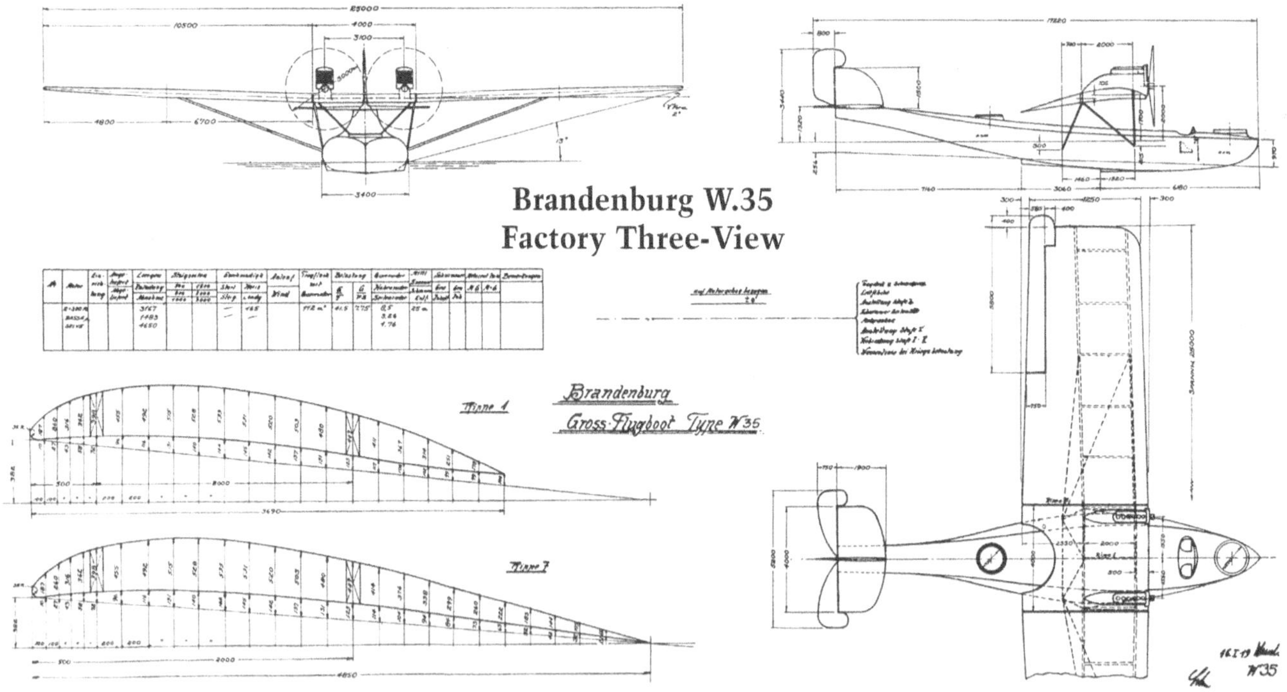

Brandenburg W.35 Factory Three-View

Brandenburg W.37

An enlarged monoplane derived from the earlier W.29 to W.34 series of floatplanes, three prototypes were ordered in August 1918. When the Navy stopped work in December only one fuselage had been completed, however, with design refinements this machine eventually appeared as the Caspar S.I post-war. *Jane's All the World's Aircraft* noted that the S.1 "is derived from the Hansa-Brandenburg sea-monoplane so well known during the war." The modifications introduced since the war were the installation of a roomy cabin for four passengers, a better mounting of the engine to facilitate inspections, a stronger undercarriage and improved float design. Instead of the 260-hp Maybach Mb.IV previously fitted, a Mercedes D.IVa of the same horsepower was fitted. The version being built for the Swedish Navy was stated to have the 240-hp Siddeley Puma engine, and one model was to be built with the Rolls-Royce Eagle IX engine.

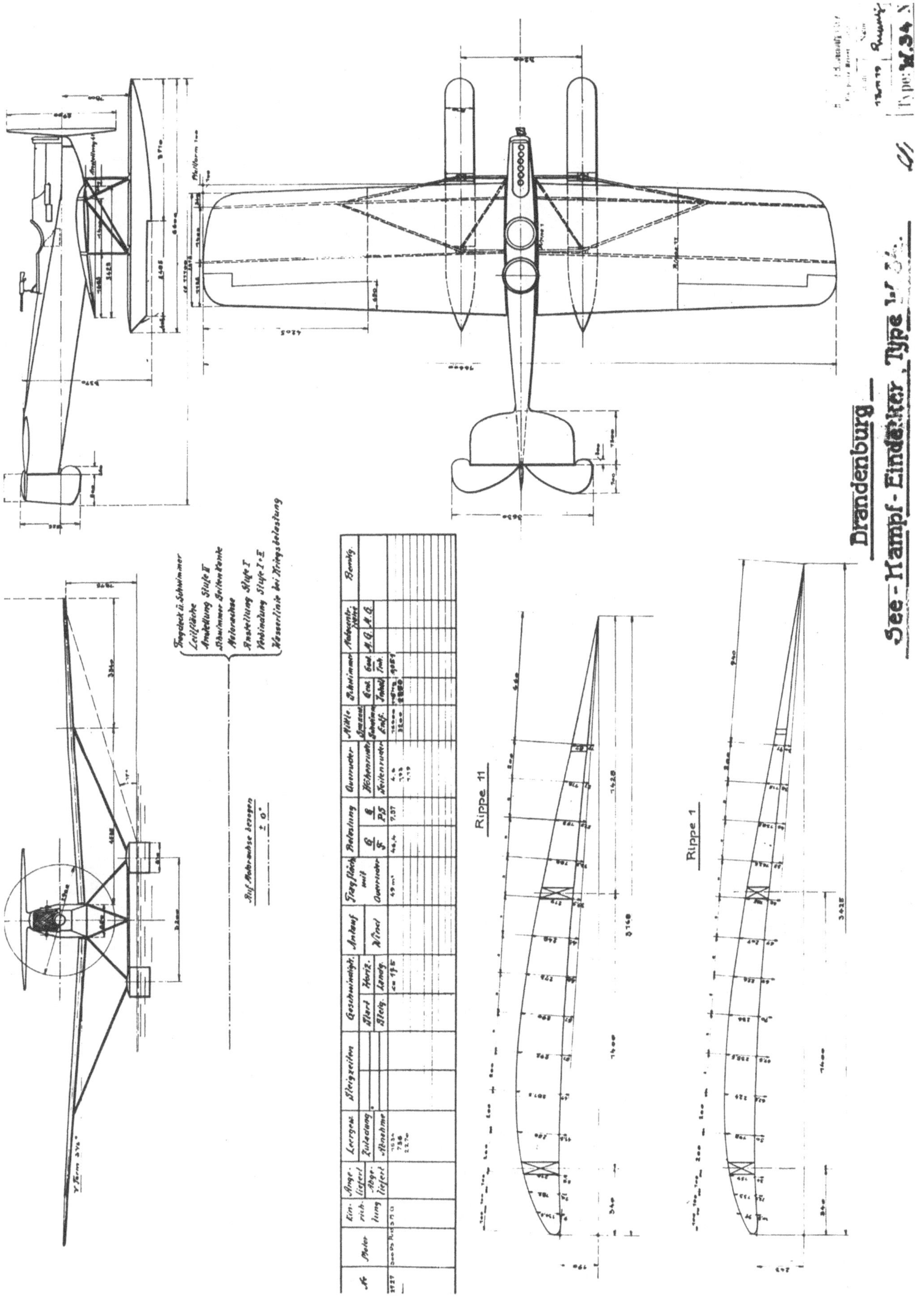
Brandenburg
See-Kampf-Eindecker Type
Type: W.34
Rippe 11
Rippe 1

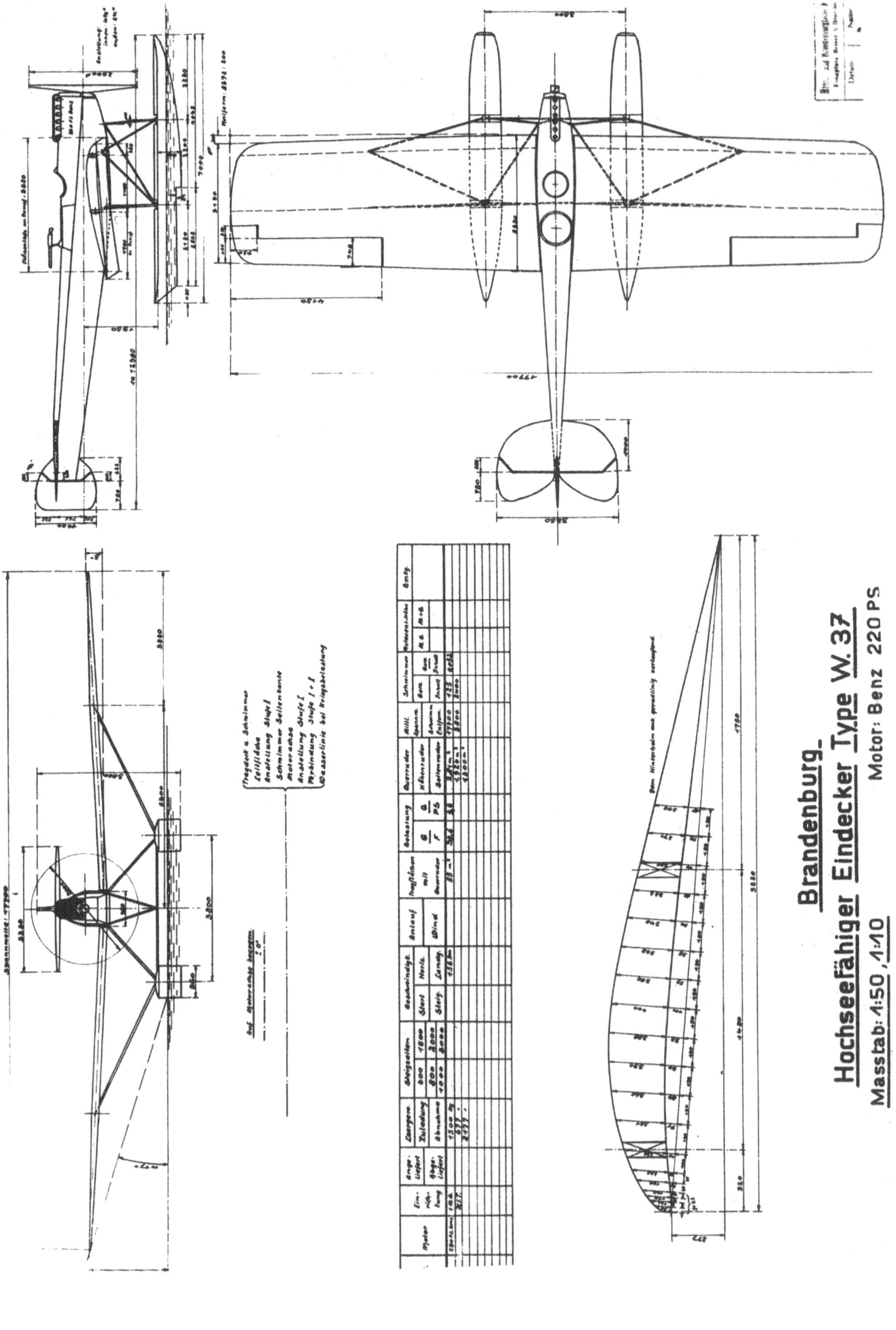
Brandenburg
Hochseefähiger Eindecker Type W.37
Masstab: 1:50, 1:10
Motor: Benz 220 PS

Caspar U 1

Above & Below: The Caspar U 1 was an all-metal aircraft designed to be carried by a submarine for reconnaissance. The U 1 was designed to be carried in a waterproof metal housing in a dismantled state and assembled on the submarine for its reconnaissance flights. It was completed postwar.

DFW Flying Boat

Above: DFW flying boat in the DFW factory; the radiator on the nose is distinctive. This one-of-a-kind 1914 design had a Daimler engine in the hull driving a pusher propeller via a drive shaft. The empty weight was 820 kg. DFW built no other seaplanes, perhaps because they owned the Lübeck-Travemünde firm which later produced a small number of floatplanes. (Peter M. Grosz Collection/SDTB)

Fokker W.1

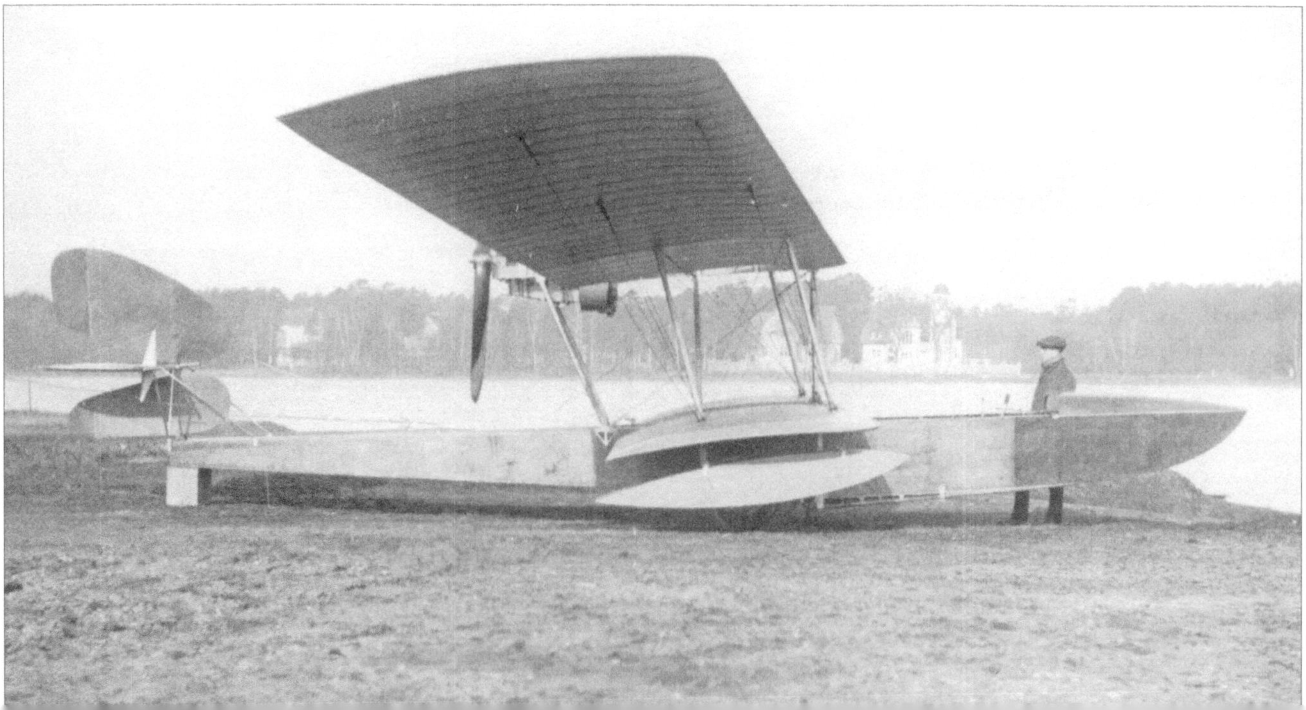

Above: The Fokker W.1 flying boat was powered by a 70 hp Renault Fokker had at hand. (Peter M. Grosz collection/STDB)

Facing Page, Below: Side view of the Fokker W.1 flying boat. (Peter M. Grosz collection/STDB)

Below: The only Fokker W.1 monoplane was built in 1913. (Peter M. Grosz collection/STDB)

Fokker W.2

Above: This photo is part of a take-off sequence. The W.2 had a tail float in addition to the main floats. Intended for a competition, the Fokker W.2 was not competitive and was scrapped. (Peter M. Grosz collection/STDB)

Fokker W.3

Above: Although the W.3 was a hydroplane, not an airplane, it is included for completeness. Components from the W.2 were included in its construction.

Fokker W.4

Above: The third Fokker seaplane was the W.4; it was a floatplane version of the M.7. Underpowered by its 80 hp Oberursel U.0, it could not take off from the water despite its large, 2-bay wings. Its interplane struts sloped inward at the top. It was converted back to a W.7 landplane. (Peter M. Grosz collection/STDB)

Fokker C.IW

Above: The Fokker C.IW was a postwar development. Fokker C.I work number 3972 was converted to a seaplane and first flew in September 1920. On 6 and 11 October 1920 it was tested on the Schweriner See. Its further fate is unknown. (Peter M. Grosz collection/STDB)

Friedrichshafen FF1

Above: Friedrichshafen's first airplane design, the primitive FF1. Appropriately, it was a seaplane as were most Friedrichshafen designs. The FF1 first flew in mid-November 1912, the year that the company was founded.

Above: The Friedrichshafen FF1 had a central main float that could be fitted with wheels to facilitate ground handling. Side-mounted radiators cooled the 100 hp Argus pusher engine and the ailerons were mounted between the wings, which had 3½ bays of interplane bracing. Friedrichshafen aircraft not shown were landplanes.

Friedrichshafen FF2

Above: The Friedrichshafen FF2 was a small, single-seat monoplane with two floats that was built for a private customer. Engine was a 50 hp Orlikon.

Friedrichshafen FF4

Above & Below: Two views of the Friedrichshafen FF4 training floatplane built for a private customer. The engine was a 65 hp Argus with brow radiator. The aircraft followed the basic *Taube* configuration so popular in pre-war Germany due to its inherent stability. Twin main floats and tail float were fitted. (Lower photo the Peter M. Bowers Collection/The Museum of Flight)

Friedrichshafen FF8

Above: The Friedrichshafen FF8 floatplane before markings were applied. A 65 hp Argus engine with side radiators was fitted, but the FF8 reverted to wing-warping for roll control. One reference states the FF8 was a single-seater, although it clearly had room for a second seat in front of the pilot. The rudder and fin were larger and the rear float strut was longer than the FF4. (The Peter M. Bowers Collection/The Museum of Flight)

Above: The Friedrichshafen FF8 floatplane after markings were applied. The ground crew appear to be in the process of launching the aircraft. Side radiators were fitted.

Friedrichshafen FF9

Above: The Friedrichshafen FF9 reverted to the configuration of the earlier FF1 and greatly resembled it, but was a larger, more powerful aircraft powered by a 135 hp NAG engine.

Above: The Friedrichshafen FF9 was a larger, more powerful aircraft than the FF1 it resembled. Here it uses the wheels that could be extended for land operations. The text on the rudder reads "F.F. Flugzeugbau Friedrichshafen G.m.b.H."

Friedrichshafen FF11

Right: The FF11 was Friedrichshafen's first flying boat. The hull was built by the Oertz yacht works and the similarity to Oertz flying boats is obvious. The similarly-powered FF9 had notably better performance than the FF11 due to the flying boat's greater size and weight. The SVK assigned Marine Number 41 to the FF11.

Right: The FF11 after its nose contours were modified

Friedrichshafen FF15

This Page: The Friedrichshafen FF15 followed the configuration of the earlier FF1 and FF9. The FF15 was basically an improved FF9 and had similar specifications and engine type – 135 hp NAG.

Friedrichshafen FF17

Above: The FF17 was a much more modern design than earlier Friedrichshafen seaplanes. It was a 2½-bay tractor biplane with side radiators. The inner struts slanted inward at the top wing. Here the FF17A has a single central float with large wingtip floats for stability on the water. The engine was a 135 hp NAG like some earlier types.

Above: The FF17 was later tested with dual main floats as the FF17B. This configuration resembled the Friedrichshafen seaplanes used during the war. (The Peter M. Bowers Collection/The Museum of Flight)

Friedrichshafen FF19

Above: The first production Friedrichshafen FF19, Marine Number 25, trails a red wingtip identification streamer.

Below: The FF19 was the first Friedrichshafen design to be produced in quantity. A total of 16 airframes were built; 15 were delivered to the Navy and one airframe was used for structural testing. The engine was a 100 hp Mercedes D.I.

Friedrichshafen FF21

Above & Below: The FF21 was developed from the Friedrichshafen FF11 and looks virtually identical in side view to the Oertz F.B.1, leading one to wonder if they were indeed the same design. Oertz built the hulls for both the FF11 and FF21, and there is a great resemblance between these designs. A 150 hp Benz Bz.III engine was mounted in the hull.

Friedrichshafen FF27

Above: The FF27 was a refinement of earlier Friedrichshafen pusher biplanes; the one built was accepted by the German Navy and assigned Marine Number 62. Wheels could be attached as shown for easier ground handling. The drag-creating pusher configuration was little used by German manufacturers after the war started. Pushers had good visibility for the crew but were too slow. Later, when air-to-air combat became more common, pushers were too vulnerable to attack from the rear. A 135 hp NAG engine was fitted. The sole FF27 was designed for the 1914 Baltic Seaplane Competition.

Friedrichshafen FF29

Above: FF29s Marine Numbers 208, 204, and 407 are lined up with Oertz F.B.3 MN 46 at Zeebrügge in early 1915. The FF29 was a 3-bay biplane floatplane. There were some variations in the design of the vertical tail surfaces.

Above: Friedrichshafen FF29 Marine Number 211 was an FF29 powered by a 120 hp Mercedes D.II.

Above: Friedrichshafen FF29A Marine Number 412 was delivered in March 1915.

Below: A Friedrichshafen FF29 being brought ashore.

Friedrichshafen FF31

Right: Friedrichshafen FF31 Marine Number 274 was one of only two FF31 aircraft built, the other being MN 275. It was a 3-bay pusher, twin-float biplane powered by a 160 hp Maybach Mb.III.

Above & Right: Friedrichshafen FF31 Marine Number 275. It was armed and a tail float was fitted. Its nose contours appear to be more rounded than those of MN 274. (photo at right courtesy Chuck Thomas)

Friedrichshafen FF33 Series

Left: Marine Number 453 was a Friedrichshafen FF33 powered by a 120 hp Mercedes D.II. It has 3 bays of bracing, two main floats, and a tail float, but was not armed.

Friedrichshafen's FF33 family was the backbone of German naval aviation during most of the war. Of the 854 seaplanes Friedrichshafen provided to the Navy, 427, or half, were from the FF33 family. Friedrichshafen's seaplanes were consistently more robust and reliable than those from competing manufacturers and had good flying and seakeeping qualities, the key reasons Friedrichshafen aircraft dominated the Navy's production orders. While combat aviators greatly appreciate sparkling performance, reliability is even more important, and that is especially true of seaplanes.

Left: Marine Number 453 was a Friedrichshafen FF33 powered by a 120 hp Mercedes D.II with side radiators.

Above: Friedrichshafen FF33D Marine Number 460 was a powered by a 160 hp Maybach Mb.III. Only two FF33D aircraft, MN 459–460, were built; they were class CB, meaning an armed two-seater. Later the class designation for an armed two-seater was shortened to class C. The FF33D was the first FF33 variant to be armed with a machine gun as shown. While assigned to Zeebrügge, on the night of 22/23 Jan. 1916, this aircraft attacked Dover with 10 bombs.

Above: Friedrichshafen FF33D Marine Number 459 ready to launch; it was the only other FF33D. The FF33B also used the 160 hp Maybach Mb.III but was class BFT, a two-seat aircraft with bombing equipment and wireless transmitter but no gun. Both FF33B and FF33D had 3 bays of bracing, twin main floats, and a tail float.

Above: Friedrichshafen FF33E Marine Number 501 on patrol. This early FF33E retains the tail float featured by earlier designs, but later production batches of the FF33E would eliminate the tail float for reduced weight and drag.

Below: Friedrichshafen FF33E Marine Number 634 on its beaching at Warnemünde; it was ordered in January 1916. The later-production FF33E eliminated the tail float and added a fin under the fuselage for greater stability. The FF33E was unarmed and, like all subsequent FF33 variants, was powered by a 150 hp Benz Bz.III. MN 510 was the only armed FF33E.

Above: A striking view of a Friedrichshafen FF33F. Unfortunately, the Marine Number to verify that is not fully visible, but there is a gun mounting for the observer. The FF33F was the armed version of the FF33E; it had the same dimensions and 150 hp Benz Bz.III engine. An airfoil radiator supplied the cooling.

Above: The observer of a Friedrichshafen FF33F passes a message to a U-boat. Again there is a gun-ring and gun mounted for the observer.

Left: Friedrichshafen FF33H Marine Number 695 flying a mission from Angersee. This aircraft, an armed class C, was the first aircraft of the 5th FF33H production batch ordered in April 1916. The red identification pennants can be seen attached to both #695 and the aircraft from which the photo was taken. As all FF33 variants from the FF33E on, the engine was a 150 hp Benz Bz.III. The profile of the fins and rudder was distinctive. An airfoil radiator was fitted to MN 695.

Above: Friedrichshafen FF33H Marine Number 596 was the first production FF33H. A class C, it was ordered in January 1916. The FF33H was a more compact design than the FF33E, having 2 bays of struts compared to the 3 bays of struts used in previous FF33 variants. An underwing radiator was fitted.

Above & Below: Friedrichshafen FF33J Marine Number 3033 was a class HFT ordered in March 1918, a remarkable achievement for a basic design that originated in 1914. Unusually for a slow floatplane, the observer has a streamlined headrest. It has a streamlined nose with propeller spinner and 3 bays of bracing.

Bottom: FF33J Marine Number 303x, a class HFT aircraft, preparing for a mission.

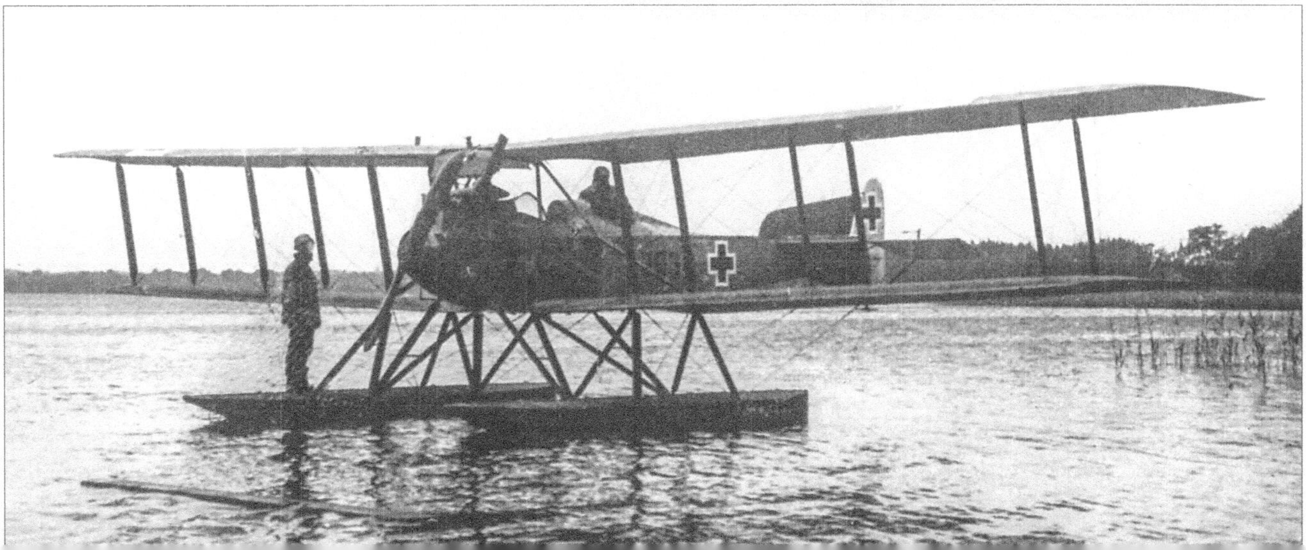

This Page: Marine Number 932 was the only Friedrichshafen FF33K built. A class C2MG, it was an intermediate step to the definitive FF33L and was built as the first aircraft of the first FF33L production batch ordered in September 1916. The 2-bay FF33L was a smaller, lighter, more refined design than the FF33K and offered better performance and maneuverability. The FF33K tail was essentially the same as that of the FF33E.

Above & Below: Marine Number 933 was the first Friedrichshafen FF33L built. A class C2MG, it was a smaller, lighter aircraft than the FF33K that was slightly faster and had double the rate of climb. It was also more maneuverable, all qualities that enhanced its effectiveness for its designed role as the Navy's first two-seat floatplane for the C2MG fighting role. Like the FF33K it had a 2-bay wing but it had a smaller, more streamlined vertical tail.

Above & Below: FF33L #1257, class CHFT, being recovered. This aircraft had a modified tail without fixed fin. From the clothing worn by the civilian onlookers the location was Turkey.

Below: FF33L #1262, class CHFT, with standard vertical tail surfaces. (The Peter M. Bowers Collection/The Museum of Flight)

Above: FF33L #1590, class C2MG, had the modified tail without fixed fin.

Below: FF33L #3144, class C2MG, had an experimental enlarged rudder; this photo shows it without a spinner. (The Peter M. Bowers Collection/The Museum of Flight)

Friedrichshafen FF34

Above & Below: The sole prototype Friedrichshafen FF34 was built to explore a more streamlined pusher configuration for reconnaissance floatplanes. It was a large aircraft and was powered by a 240 hp Maybach Mb.IV engine. It was assigned MN 117.

Friedrichshafen FF35

Above: Front quarter view of FF35 Marine Number 300 shows its basic configuration as a twin-engine floatplane. The 160 hp Mercedes D.III engines are mounted as pushers in primitive nacelles with large, frontal radiators. The horizontal tail was mounted well above the fuselage to keep it out of the water spray from the floats. This was the first Friedrichshafen torpedo plane built; it remained a single prototype; a second planned aircraft was not built. The sole FF35 built was sent to Angernsee in the Baltic and was used for bombing.

Above: Side view of the FF35 prototype torpedo bomber on its beaching dolly outside the assembly hall.

Friedrichshafen FF39

Above: Marine Number 587 ordered in December 1915 was the prototype FF39. Powered by a 200 hp Benz Bz.IV, the FF39 was a class CHFT floatplane. It was a 3-bay, twin-float biplane with streamlined nose and tail surfaces similar to the other FF33E. A total of 14 of these aircraft were built. Ailerons were fitted to the upper wing only. (The Peter M. Bowers Collection/The Museum of Flight)

Above: Ordered in December 1915, first flight of the prototype FF39 was in December 1916 and it was delivered about 25 March 1917. The prolonged development time was probably due to delayed availability of the Benz Bz.IV engine. (The Peter M. Bowers Collection/The Museum of Flight)

Above: In 1918 FF39 #645, with *Ltn.z.S.* Wolfram Eisenlohr as observer, rendezvouses with a submarine one of Eisenlohr's brothers was serving aboard. Eisenlohr sank the Russian destroyer *Stroini* with bombs while flying FF41AT MN 1000. MN #645 had a more pointed spinner than 587.

Below: FF39 #644 with no spinner wears the late-war camouflage. (The Peter M. Bowers Collection/The Museum of Flight)

Above & Below: FF39 Marine Number 1127, the first aircraft of the main production batch, wears an interesting, non-standard camouflage scheme. These photos show it with no spinner. The vertical tail is refined compared to the prototype. This production batch had ailerons on all wings connected by actuating rods. (Top photo The Peter M. Bowers Collection/The Museum of Flight)

Friedrichshafen FF40

Above & Below: FF40 Marine Number 669 shows its unusual configuration of a single engine driving two propellers via ZF bevel gears and propeller shafts, which were enclosed in streamlined fairings. The FF40 was relatively large and powerful for a two-seat reconnaissance floatplane, but the power transmission system was heavy, limiting its equipment to a flexible gun and a wireless transmitter only with no receiver. Wing bracing was 3 bays and two main floats were fitted with no tail float. (Top photo The Peter M. Bowers Collection/The Museum of Flight)

Above: The single 240 hp Maybach Mb.IV engine in the FF40 drove two propellers via ZF bevel gears and propeller shafts. The FF40 was ordered in March 1916 and, despite its relatively complex power system, was quickly built and flown. First flight is reported to be in April 1916 with delivery to the SVK in August 1916.

Friedrichshafen FF41A & FF41AT

Above: The sole FF41A prototype torpedo bomber on its beaching dolly outside the assembly hall. The FF41A had a central fin and three rudders; the production FF41AT had a simplified tail with a single fin and rudder.

Facing Page, Below: Production FF41AT MN996 torpedo bomber with streamlined engine nacelles.

Above: FF41A Marine Number 678 torpedo bomber with torpedo under test before the engine nacelles were fitted.

Above: This aircraft is neither FF41A Marine Number 678 nor a later FF41AT; instead, it may be an intermediate design. Might this be Marine Number 735, the Friedrichshafen torpedo bomber that was cancelled?

Above: Marine Number 996, the first FF41AT torpedo bomber built, on its beaching dolly.

Above: Marine Number 997, the second FF41AT torpedo bomber built, is being eased into the water. The production FF41AT aircraft were flown extensively on operations in the Baltic, but these missions were primarily mine laying, conventional bombing, and maritime reconnaissance, not torpedo bombing.

Friedrichshafen FF43

Above: The FF43 under evaluation at the *Seeflug-Erprobung* (Seaplane Test Facility) at Warnemünde. With better speed than the Albatros W4, the FF43 seemed to offer as much potential as the W4. However, the FF43 was not chosen for production, perhaps due to a mediocre rate of climb. One gun was fitted.

Above: The FF43 prototype with Marine Number and national insignia applied. Despite remaining a single prototype, the FF43 was assigned to Zeebrugge for operational service. On November 10, 1916, *Lt.z.S.* Schuler shot down one of two Short seaplanes he encountered during a coastal patrol between Zeebrugge and Dunkirk while flying the FF43. This was probably Short 184 serial 9054 of Ft.Lt. JD Hume & CPO WE Bradley from Westgate who were KIA in North Sea that day.

Friedrichshafen FF44

Above & Below: Rebuilt from the FF34 pusher, the FF44 featured a conventional 3-bay configuration and retained the Marine Number of 117 first used by the FF34. The FF44 followed the basic Friedrichshafen floatplane formula except for its vertical tail surfaces, which have no fixed fins, just a balanced rudder. This design could also be found on some FF33L aircraft.

Friedrichshafen FF48

Above & Below: The Friedrichshafen FF48, MN 1472, was a larger aircraft in the class of the W19 and like the W19 was powered by the 240 hp Maybach Mb.IVa. The Brandenburg W.19 was chosen for production. (The Peter M. Bowers Collection/The Museum of Flight)

Friedrichshafen FF49

Above: FF49C Marine Number 1521 was the first Friedrichshafen FF49 built. The FF49C was a more powerful development of the FF33 series to perform the same tasks with better performance and heavier payload. Balanced upper ailerons reduced the pilot's workload on patrols while ailerons on all wings provided better maneuverability.

Above: FF49C #1872 was the first aircraft of a batch of 30 (1872–1901) ordered from Sablatnig in August 1917. The FF49 was powered by the 200 hp Benz Bz.IV engine. Built with 3-bay bracing, later production aircraft had ailerons on all wings with the ailerons on each side connected by an actuating strut.

Friedrichshafen FF53

Above: Marine Number 1663 was the first of three FF53 torpedo bombers ordered in June 1917. Because SVK records are incomplete after June 1918, the delivery dates for these aircraft are not known. In fact, this photo is the proof the FF53 was actually built and it is not certain the last two aircraft ordered were completed. The FF53 was powered by a pair of 260 hp Mercedes D.IVa engines and greatly resembled the Friedrichshafen G.IIIa land bomber, but the engines were mounted in tractor configuration to keep the propellers out of the spray of water from the floats. Unfortunately, almost nothing is known of the aircraft's specifications or performance and there is no record of its operational use. Designed as a torpedo bomber, the aircraft arrived after torpedo dropping had been abandoned as an operational role, and torpedo bombers had been re-assigned to long-range maritime reconnaissance and bombing.

Friedrichshafen FF59

Above: The FF59 was developed from the FF49 with particular attention to giving the observer the maximum field of fire. Powered by a 200 hp Benz Bz.IV, it was a 2-bay design with streamlined nose and propeller spinner.

Above: Marine Number 1822 was the first prototype FF59. The vertical tail surfaces were mounted under the tailplane to maximize the observer's field of fire in a manner reminiscent of Brandenburg practice. Developed from the FF49, the FF59 had the same engine but 2-bay instead of 3-bay bracing. (The Peter M. Bowers Collection/The Museum of Flight)

Above: This photo shows Marine Number 1822 after the vertical tail surfaces were modified. Now the fin projects above the fuselage to improve flying qualities but compromising the observer's field of fire.

Friedrichshafen FF60

Above: The FF60 during testing on Lake Constance. The FF60 was completed just as the war ended but Friedrichshafen was given permission to test fly the aircraft. Designed for long-range maritime reconnaissance, it had a unique triplane configuration with its four 170 hp Mercedes D.IIIa engines mounted on the lower and middle wings.

Below: The FF60 being launched via crane for testing on Lake Constance.

Friedrichshafen FF63

This Page: Powered by the 200 hp Benz Bz.IV, the Friedrichshafen FF63 appears to be a fair design that just does not match the innovative structural design of the W29 and W33 that were its competitors. The additional bracing struts above the wing and the struts above the fuselage added weight and drag that the Brandenburg monoplanes avoided. The tail was made of fabric-covered steel tubing, a major innovation for Friedrichshafen. The drag-producing radiator mounted over the engine was undoubtedly fitted to expedite flight testing and not intended for production.

Friedrichshafen FF64

Above: Designed for shipboard use, the prototype FF64 shows careful streamlining and robust construction for its time.

Above: This is believed to be the first version of the prototype FF64 tail design with no fin on top of the fuselage. The wings folder for shipboard storage.

Below: Believed to be the second configuration of the prototype FF64; the tail has an upper fin and the lower fin has been shortened.

Above: The FF64 was designed to carry a flexible machine gun and wireless to lead a commerce raider to target ships. The tail design was influenced by Brandenburg designs to give the observer a wide field of fire.

Third tail variation?

Friedrichshafen FF67

Above: The FF67 was a postwar transport floatplane based on the FF49. The engine was a more powerful 260 hp Mercedes D.IVa engine and the tail surfaces were modified. Here it is seen after the passenger cabin had been added in March 1919. (Peter M. Bowers Collection, Museum of Flight)

Above: Another view of the sole protottpe FF67 after the passenger cabin had been added.

Friedrichshafen FF71 & FF71a

Above: The postwar Friedrichshafen FF71 transport floatplane as originally built with open cockpits. It still carries the camouflage fabric used on warplanes. It was Friedrichshafen's second attempt to create a passenger floatplane based on the FF49. The engine remained the 200 hp Benz Bz.IV fitted in the FF49 and the vertical tail surfaces were larger than the FF67.

Above: The Friedrichshafen FF71 transport floatplane after modification to have a passenger cabin. The pilot sat in an open cockpit; this helped the pilot sense the flight performance of the floatplane in light of the modest instrumentation. it is alleged that five FF71a transports were later built by LFG. A side drawing of the FF71a shows it had a nose radiator replacing the leading edge radiator.

Gotha *Büchner See*

Above: The Büchner *See* pusher floatplane was the first Gotha seaplane. Built in 1913, it was powered by a 100 hp Mercedes D.I. Only one was built, but it started Gotha in the seaplane business.

Gotha WD1

Above: The prototype WD1 was powered by a two-row, 100 hp Gnôme rotary engine. The WD1 floatplane, MN 59, was a conventional 3-bay biplane; a small float was attached to the rudder. "WD" stood for "*Wasser Doppeldecker*" (water biplane).

Above : The six subsequent WD1 floatplanes were powered by a 100 hp Mercedes D.I engine. This WD1 is almost certainly the second WD1 built by Gotha for the *Ostseeflug Warnemünde* 1914 competition. It has the Gotha name painted on the fuselage and retains the small float attached to the rudder. This aircraft was later purchased by the Navy.

Gotha WD2

Above: The first WD2 floatplane was powered by a 150 hp Rapp engine and was entered in the *Ostseeflug Warnemünde* 1914 competition as competition #19a. It has the Gotha name painted on the fuselage; this aircraft was later purchased by the Navy and assigned Marine Number 61. The first WD2 was basically a WD1 with different engine.

Above: WD2 Marine Number 236 is seen taxiing for take-off. The float attached to the rudder of the early WD1 aircraft was no longer used on subsequent types and its comma-shaped rudder without fixed fin has been replaced with fin and revised rudder. Its 160 hp Mercedes D.III gave it much more power than the WD1. At least three WD2 floatplanes were sent to the German *Wasserfliegerabteilung* in Turkey in October 1915. In addition, at least ten WD2 floatplanes were delivered to Turkish forces.

Above: WD2 MN 424 (shown here) and 425 were fitted with temporary wheels and tail skids attached to their floats to enable overland flights to Turkey because rail shipments through Romania were banned. These two aircraft were powered by the 160 hp Mercedes D.III engine and had a gun turret for the observer mounted above the upper wing, giving the gunner a 360° field of fire. The fin was greatly enlarged compared to unarmed MN 236 to counteract the destabilizing effect of the gun turret and standing gunner. WD2 MN 424 was at Chanak when credited with a victory in Jan. 1916 by *Flgobmt* Wilhelm Schubert (pilot)/*Flgmt* Werdier (observer), who shot down a Farman that crashed east of Tenedos. Later this aircraft was at Kawak.

Facing Page: Detailed view of the gun turret installation on a WD2.

Gotha

Gotha WD3

Above: The Gotha WD3 is seen in the Gotha factory pond before its Marine Number 259 was applied. Note its center fin and insignia on top of the lower wing. It has 2 bays of struts outboard of the tail booms and twin floats with no tail float. The sole Gotha WD3 later went to Stralsund/Wiek as a training aircraft, being flown there on 2 August 1916.

Above: The WD3 floatplane in the Gotha factory pond before its Marine Number 259 was applied. The WD3 floatplane had three crew members, the pilot aft and two observers forward, an unusual arrangement; most similar aircraft had only one observer.

Gotha WD4 / also known as UWD

Above: The Gotha UWD attracted a crowd during its sea trials in February 1916. So many people rode in it that it was nicknamed the "Trojan Horse". UWD stood for Ursinus water biplane; Oskar Ursinus, editor of *Flugsport*, was the designer. Power was provided by two 160 hp Mercedes D.III engines.

Below: The Gotha UWD in flight displays its distinctive lines and the clear field of fire for its gunners, especially the nose gunner. The second gunner sat in the middle with the pilot aft as a result of its 'battle-plane' heritage from the Gotha G.I. MN 120 was assigned to the only UWD built. Assigned to Zeebrugge, it flew on operations and bombed the UK.

Gotha WD5

Above: The Gotha WD5 in the Gotha factory pond; MN 118 was already applied. The compact, 2-bay design was for more speed; power was from a 160 hp Mercedes D.III engine. The insignia are on both top and bottom surfaces of the top wing. Intended for bombing, the sole WD5 was sent to Turkey on 13 July 1916.

Above: The Gotha WD5 in the Gotha factory pond shows its comparatively compact, 2-bay design.

Gotha WD6

Above: The Brandenburg NW was built under license by Gotha as the Brandenburg NW(Go). Gotha built 30 of these floatplanes that were given the internal Gotha company designation WD6.

Above: Brandenburg NW(Go) Marine #759 was destroyed landing at Libau on 15 June 1917.

Gotha WD7

Above: The prototype Gotha WD7, Marine #119, in the Gotha factory pond. The WD7 had distinctive tail surfaces with three fins and two rudders with aerodynamic balances at the bottom. The WD7 was designed as a 3-seat *Kampfflugzeug* meant to destroy enemy aircraft, but by the time the production batch was available that concept was out-moded.

Above: Gotha WD7 Marine Number 671 afloat in the Gotha factory pond. Power was two 100 hp Mercedes D.Is.

Gotha WD8

Above: Gotha WD8 Marine Number 476 afloat in the Gotha factory pond. The WD8 used the same wing cellule, tail surfaces, and floats as the WD7 and was built as a comparison aircraft to the WD7 because both aircraft had the same horsepower. Confusingly, the wrong Marine Number was painted on the floatplane when these photos were taken, a mistake later corrected. A 240 hp Maybach Mb.IVa engine was installed.

Above: Side view of Gotha WD8 Marine Number 476 afloat in the Gotha factory pond. The wrong Marine Number was painted on the WD8 when these photos were taken. With the same total engine power as the WD7 had from two engines, the WD8 was 10 kmh faster due to lower drag.

Gotha WD9

Above: Gotha WD9 Marine Number 572 on beaching dollies. The inner wing bay wire bracing of the WD9 was eliminated to enable the gunner to safely fire forward between the wings. The engine was a 160 hp Mercedes D.III. Only one WD9 was built. Initially assigned to *Flandern* I at Zeebrügge, it was later sent to Turkey.

Abpve: Gotha WD9 MN 572 on beaching dollies. The WD9 had 2 bays of wing bracing and an aerodynamic balance on the bottom of the rudder, a feature that by now was becoming a hallmark of Gotha floatplanes.

Gotha WD10 / U-1

Above: The Ursinus floatplane fighter, Marine Number 782, seen from the side with floats extended. The cockpit was well aft because the 150 hp Benz Bz.III engine was set back to the center of gravity for improved maneuverability (due to reduced moment of inertia) and drove the propeller via an extension shaft. The WD10, also known as the U-1, was the second seaplane design by Oskar Ursinus, editor of *Flugsport*, that was built by Gotha.

Above: The Ursinus floatplane fighter was unique in having retractable floats, a significant innovation for its time. The inside upper edges of the floats were beveled to fit closely against the fuselage, which was shaped to accommodate them when retracted. One prototype was built.

Gotha WD11

Above: Designed as a torpedo bomber, the Gotha WD11 was a large, 3-bay floatplane powered by two 160 hp Mercedes D.III engines. This view of a WD11 emphasizes its distinctive Gotha appearance with twin fins and rudders with underslung aerodynamic balances. A form of aileron servo is mounted on the outer rear interplane struts to reduce the aileron control forces for the pilot, making this large aircraft easier to maneuver.

Facing Page, Below: Gotha WD11 MN 1376 of the third production batch was ordered in February 1917 and was attached to the *I. Torpedoflugzeug Staffel* in Windau. The torpedoman had elaborate aiming gear and a flexible machine gun. Signal flares are inserted in a belt external to the front gunner's cockpit. The radiators and not streamlined.

Above: Side view of the prototype WD11, Marine Number 679, afloat in the Gotha factory pond.

Above: Gotha WD11 showing the Whitehead G/125 torpedo hung just below the semi-enclosed fuselage cavity.

Gotha WD12 & WD12a

Right: The prototype WD12 showing the observer's faired headrest, consistent with this aircraft being unarmed. Designed for maximum speed, it was powered by a 160 hp Mercedes D.III engine. The prototype WD12 had to be modified to the WD12a before the *SVK* would accept it. The WD12 was a 2-bay biplane with twin floats and no tail float. The rudder had an underslung horn balance in characteristic Gotha style.

Above: One of the nine Turkish WD12 seaplanes showing the aft gunner's cockpit with gun ring. This distinguishes the armed Turkish WD12s from the unarmed prototype used by the German Navy. Six of the nine aircraft for Turkey had the 160 hp Mercedes; the other three had a 150 hp Benz Bz.III. The WD12 had a small spinner.

Gotha WD13

Above: This WD13 in Turkish markings on a beaching dolly was powered by a 150 hp Benz Bz.III. The observer's gun ring is clearly visible. Unlike most preceding Gotha seaplanes, the rudder has no aerodynamic balance. The fin retained its characteristic Gotha profile.

Above: The WD13 reverted to a 3-bay wing cellule. The German Navy evaluated the WD13 but did not purchase any; all WD13s were purchased and operated by Turkey. The spinner was more pointed than that of the WD12.

Gotha WD14

Above: The prototype WD14, Marine Number 801, on a beaching dolly upon delivery to the *SVK* on 16 January 1917. This aircraft has a rear gunner but no nose turret; the pilot and observer sat side-by-side. The WD14 was larger and more powerful than the WD11; the engines were two 220 hp Benz Bz.IV engines. The prototype had no nose turret and small fins and rudders. Production models had nose turrets and larger fins and rudders. Some late production aircraft had a third rudder and horn-balanced ailerons with ailerons on all wings.

Above: WD14 Marine Number 1946 of the last series was photographed after modification on 14 June 1918. Twenty-five WD14s, Marine Numbers 1946–1970, were ordered in October 1917 equipped for long-range reconnaissance with a droppable fuel tank replacing the torpedo. For improved controllability this series had an additional central rudder. Reported "totally unsuited for the front" due to poor aileron and rudder response, the modified Marine Number 1946 is shown with balanced ailerons of increased area. Only nine WD14s of this series were accepted before the Armistice. The German naval camouflage scheme standardized in April 1917 specified three-color hexagonal camouflage on all upper surfaces, with side surfaces blue-gray and undersurfaces light gray. Despite that, photos show that production models of the Gotha WD14 wore the naval lozenge fabric over all surfaces except the floats, giving them a much darker appearance than the standard camouflage.

Above: WD14 Marine Number 1946 of the last series as photographed after modification on 14 June 1918. Equipped for long-range reconnaissance with a droppable fuel tank replacing the torpedo, the additional central rudder is just visible.

Above: This appears to be WD14 Marine Number 1946. This view clearly shows the additional central rudder added to the last series to improve control response. The other rudders were also enlarged. To improve stability the fins were enlarged. The new insignia was standardized 30 March 1918. WD14 #1946 was transferred from Warnemunde to Wiek or Bug on 2 October 1918.

Gotha WD15

Above: The first prototype WD15, Marine Number 842, in the Gotha factory pond before the Marine Number was added. The design was a conventional 2-bay floatplane.

Above: The second prototype WD15, Marine Number 843, afloat in the Gotha factory pond. The second prototype WD15 incorporated a number of improvements as a result of flight testing the first prototype; the most visible was the use of ailerons on all wings, with the upper and lower ailerons connected by an actuating strut. Like the first prototype, the second prototype also suffered a number of engine problems with its 260 hp Mercedes D.IVa engine, and despite being accepted in January 1918, the *SVK* did not approve series production.

Gotha WD20

Above: Gotha WD 20 Marine Number 1515 in the factory before delivery in May 1918. Power was two 260 hp Mercedes D.IVa engines. Gotha WD 20 Marine Number 1517 had an additional, central rudder similar to that fitted to late-production WD14s. The WD20 was not accepted for operational service due to poor flying characteristics because it was tail heavy, problems similar to those plaguing the contemporary Gotha G.IV and G.V bombers.

Below: Wreck of a Gotha WD20. Three WD20 floatplanes were built as basically enlarged versions of the WD14 powered by the 260 hp Mercedes D.IVa.

Gotha WD22

Above: The first WD22, Marine Number 2133, photographed at the SVK on 3 July 1918 in its original form. The tractor engines were 160 hp Mercedes D.III engines and the pushers were 100 hp Mercedes D.I engines.

Above: The massive WD22 would seem to deserve four 260 hp Mercedes D.IVa engines of the type used in the Gotha G.IV and G.V instead of the smaller, less powerful engines fitted. That less powerful engines were installed was likely a result of the Royal Navy's Distant Blockade that was strangling German access to resources.

Gotha WD27

Above: The first WD27, Marine Number 4326, in the Gotha factory pond. The aircraft was powered by four 170 hp Mercedes D.IIIa engines. The wing cellule had 4 bays of bracing outboard of the engines. The vertical fins and rudders were much more square in shape than most Gotha floatplanes.

Above: The first WD27, Marine Number 4326, on a beaching trolley. From right, the men are: *Ing.* A. Klaube, *Chefkonstrukteur Dipl. Ing.* K. Rösner, *Major a.D.E.* Blattmann, *Geh.Kom.Rat* A. Kandt, *Ob.Ing.* G. Rhein, *Ing.* F. Wagner. The rare Gotha G.X light bomber is at left.

Halberstadt C.V Floatplane

Above: Halberstadt C.V modified into a floatplane. (Peter M. Grosz collection/STDB)

Below: Halberstadt C.V modified into a floatplane undergoing maintenance in postwar Latvian service. (Peter M. Grosz collection/STDB)

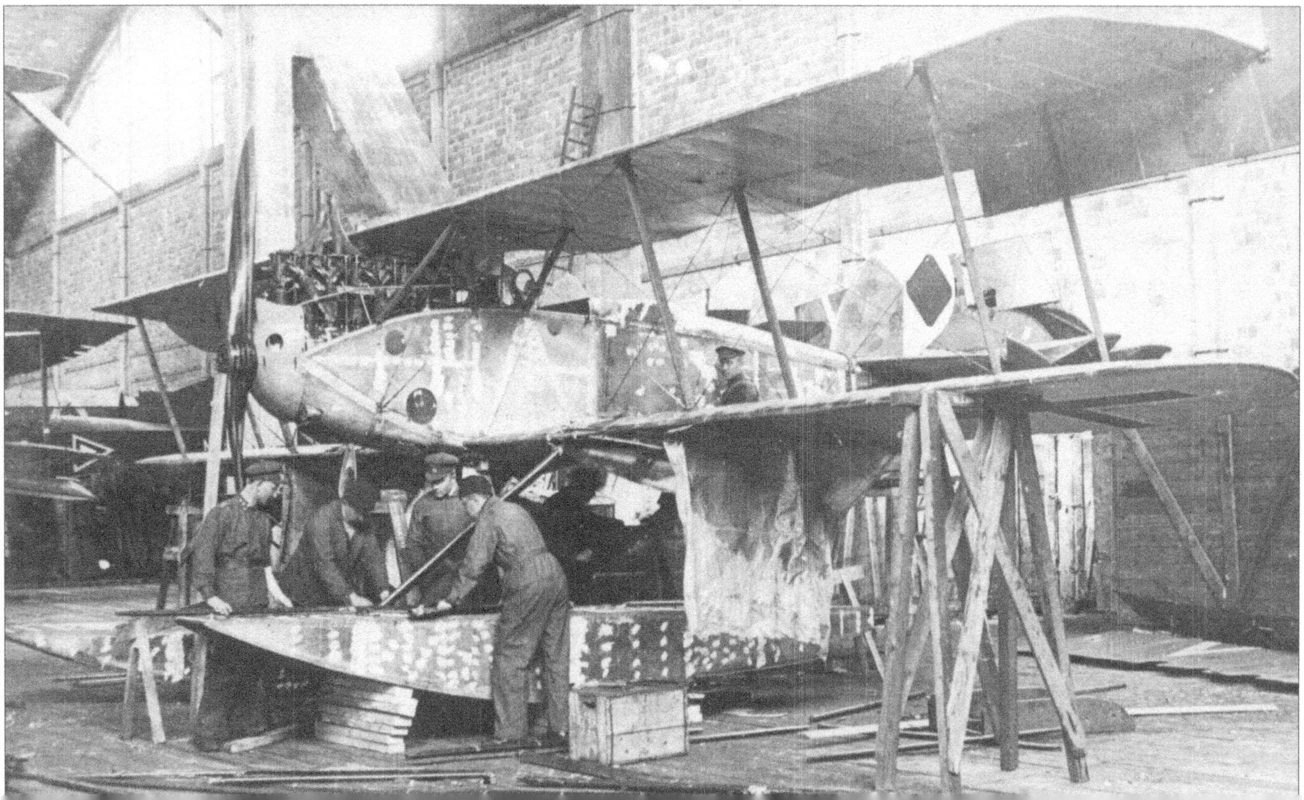

Junkers Cs.I

Above & Below: Junkers Cs.I Marine Number 7501 has the number in the Army's style rather than the Navy's. It was a floatplane derivative of the Junkers CL.I. This is the first of the three at the Seaplane Testing station at Warnemünde. Richard Thiedemann, the naval pilot who test flew the machine, was impressed with its performance.

Above: The Junkers Marine Cs.I, Marine Nummer 7503, probably at Dessau; note the camp ovens in the background. This machine has the new fin and balanced rudder. The wings have been overpainted, possibly to reduce the effect of salt water on the airframe. The fuselage national marking has not been overpainted.

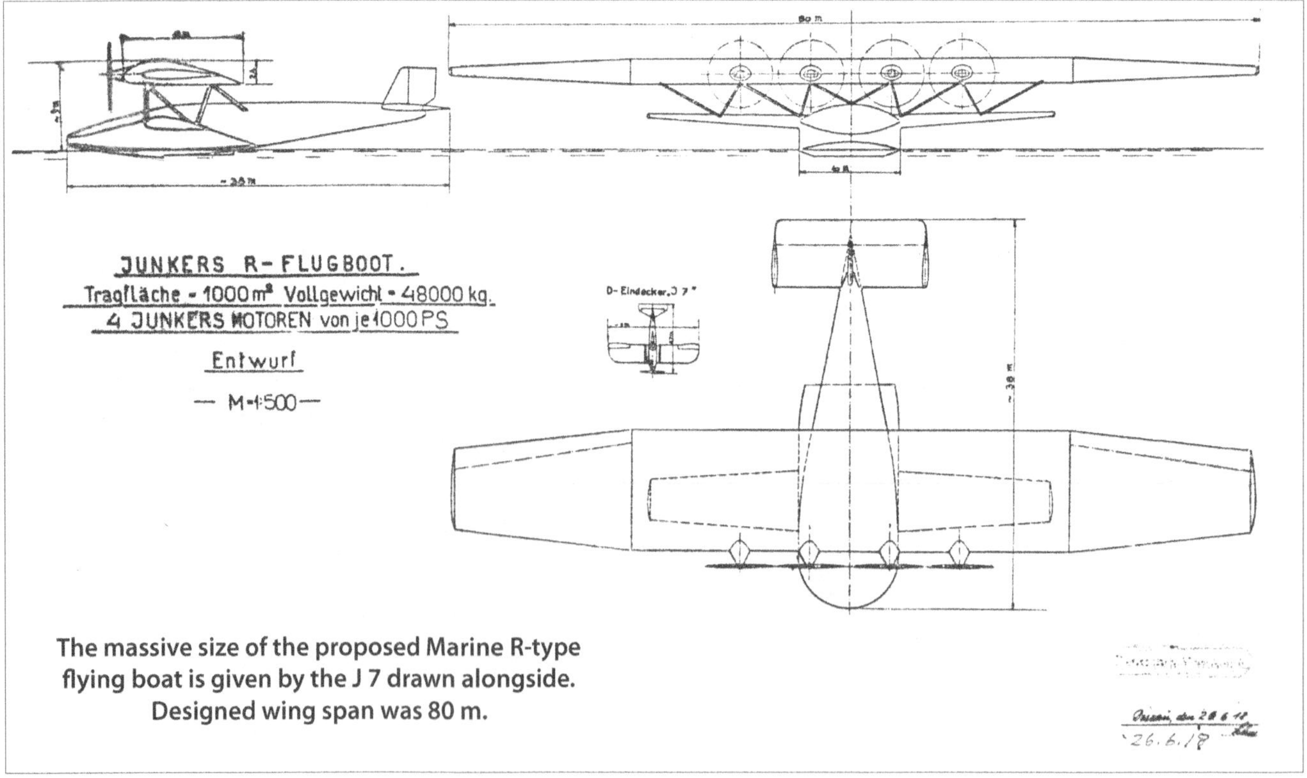

The massive size of the proposed Marine R-type flying boat is given by the J 7 drawn alongside. Designed wing span was 80 m.

Kaiserliche Werften Type 401

Above: Marine #403 ready for flight with naval pennants trailing the lower wings. It had 3 bays of bracing and a tail float. Power was from a 100 hp Mercedes D.I.

Above: Marine Numbers 404 and 405 were built by *Kaiserliche Werft* Danzig to the basic design of the *Kaiserliche Werft* Type 401 although there were likely detail differences between the Wilhelmshaven-built and Danzig-built aircraft.

Kaiserliche Werften Type 462

Above: Marine #461 on a beaching dolly; its interesting cabane and interplane bracing differed significantly from that used by Marine #462 and #463, which followed the SVK drawing of the type. It was built by *Kaiserliche Werft* Wilhelmshaven. The 2½-bay bracing was distinctive. Power was a 150 hp Benz Bz.III.

Below: Marine Number 462 was one of two built by *Kaiserliche Werft* Wilhelmshaven; the four aircraft built by *Kaiserliche Werft* Kiel to similar design differed in detail. It had normal 2-bay bracing.

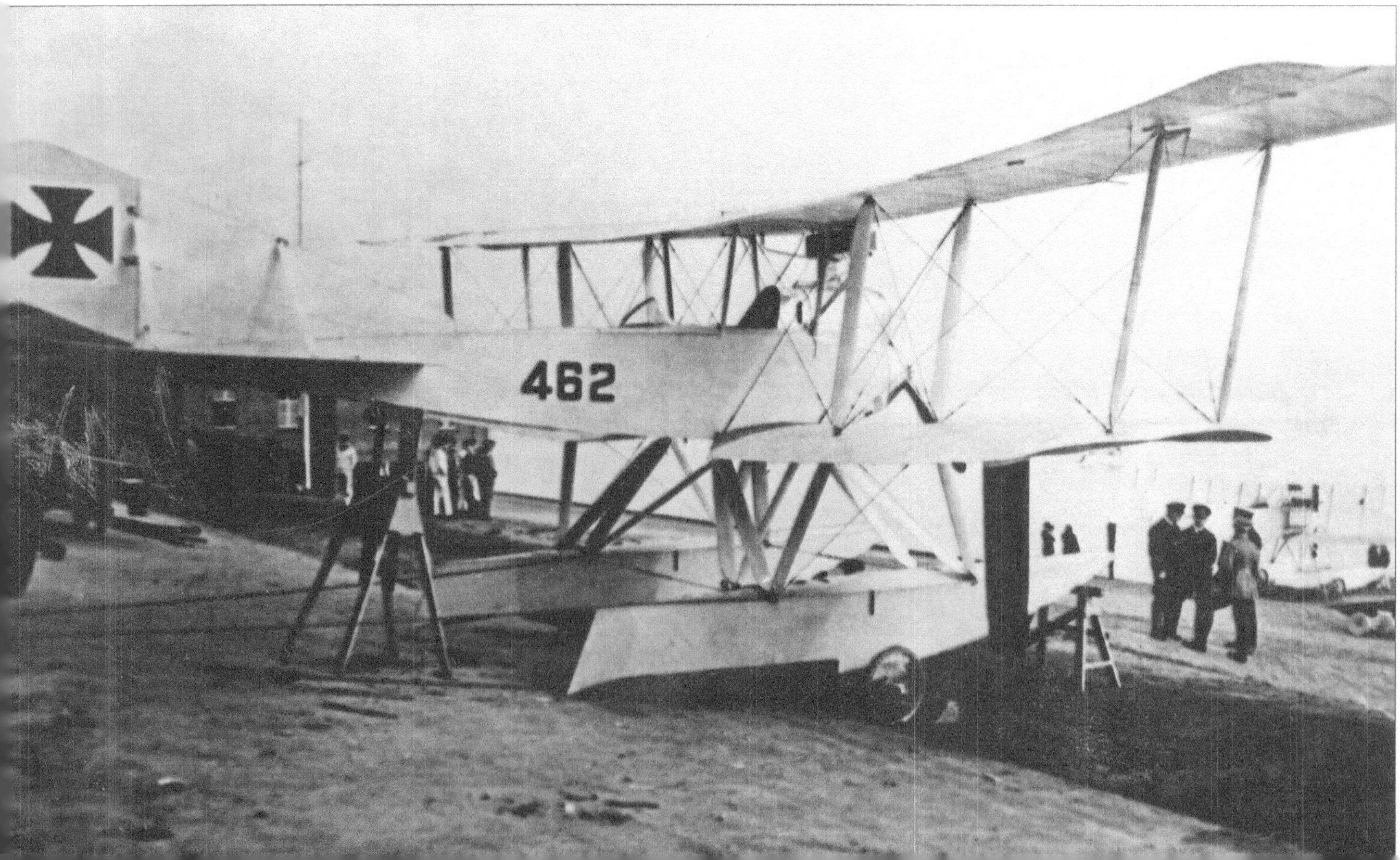

Above & Below: Marine #463, carrying standard late-war naval camouflage, shows its clean lines for a float biplane. The four Kiel-built aircraft were delivered to the naval air station at Kiel-Holtenau. The type was also known as W10. The engine was a 150 hp Benz Bz.III.

Kaiserliche Werften Type 945

Above: The KW Type 945 was a two-seat floatplane fighter category C2MG, designated 'W9' by the *Kaiserliche Werften*. Designed as a competitor to the W.12 and influenced by it, it was powered by a 150 hp Benz Bz.III. It remained a sole prototype.

Kaiserliche Werften Type 947

Above: The KW Type 947 was a two-seat reconnaissance aircraft in category CHFT. It was powered by the rare 220 hp Mercedes D.IV straight-eight engine. Evaluated by the SVK, it was not accepted. The 4-bladed propeller and large spinner were distinctive, as were the slated, 2-bay struts.

Kaiserliche Werften Type 1105

Above & Below: The KW Type 1105 was a two-seat trainer, category B/S. Evaluated by the *SVK* in 1917, it was not accepted and was returned to the manufacturer on December 7, 1917. The engine was a 150 hp Benz Bz.III.

L.F.G. Roland W1

Above & Below: The LFG Roland W1 was basically an Albatros C.Ia fitted with floats. Powered by a 150 hp Benz Bz.III, only a single aircraft, Marine Number 509, was built.

L.F.G. Roland W16

Right: The LFG Roland W16 photographed without floats in front of the LFG Bitterfeld seaplane production facility.

L.F.G. Roland W

Above: The Roland W single-seat fighter at the L.F.G. factory. It competed with the Albatros W.4, Brandenburg KDW, Friedrichshafen FF43, and Sablatnig SF4. Powered by a 160 hp Mercedes D.III and remained a sole prototype.

L.F.G. Roland V.19

Above: The all-metal LFG V.19 seaplane was a remarkable achievement for a company with no previous experience with all-metal aircraft. Not only was the airframe all metal, it was successfully designed to be stored dismantled and quickly assembled at sea for reconnaissance flights from submarines. The engine was a 110 hp Oberursel Ur.II. Postwar it was named "Putbus" after the town where it was built.

Above: The LFG V.19 was an all-metal seaplane intended for use from submarines and small surface craft. As a result it was designed to be stored dismantled in five waterproof cylinders and assembled after the submarine surfaced.

LTG FD 1

Above & Below: The FD 1 marine number 1299 floatplane fighter was powered by a 150 hp Benz Bz.III driving the propeller via an extension shaft and a reduction gear, accounting for its counter-clockwise rotation as seen by the pilot. Marine number 1299 was the first of three prototypes in the initial configuration and was eventually destroyed in structural testing. (Peter M. Grosz collection/STDB)

Above: Marine number 1518 was the first of three prototypes of the second configuration of the FD 1 floatplane fighter. Like the first configuration it was powered by a 150 hp Benz Bz.III driving the propeller via an extension shaft and reduction gear. Marine number 1518 had an enlarged fin and rudder for improved directional stability compared to the initial configuration. (Peter M. Grosz collection/STDB)

Above: Marine number 1518 was the first of three prototypes of the second configuration of the FD 1 floatplane fighter. This batch was delivered to the Navy in 1918, after the great superiority of two-seat floatplane fighters for maritime operations had been proven by the Brandenburg W.12, so there was no chance of production. (Peter M. Grosz collection/STDB)

Lübeck-Travemünde F1

Above: The F1 was the first Lübeck-Travemünde design built. Three aircraft were ordered but only the first, Marine Number 282 seen here, was built; the other two were cancelled. The F1 was a large 4-bay biplane for only 160 hp provided by its Mercedes D.III engine and that may have been a key reason only one was built.

Above: Lübeck-Travemünde F1, marine number 282, was the first aircraft built by the company. This large floatplane was powered by a 160 hp Mercedes D.III. That was not a lot of power for an aircraft of this size..(Peter M. Grosz collection/STDB)

Lübeck-Travemünde F2

Above & Below: Marine number 677 was the prototype Lübeck-Travemünde F2. The prototype was a 4-bay biplane; production F2 aircraft were 3-bay biplanes. Powered by the 220 hp Mercedes D.IV straight-eight engine, the F2 was a massive type CHFT, meaning it had a wireless transmitter and receiver and a flexible machine gun for the observer. Its four-bladed propeller was required to absorb the power of the slow-turning staight-eight engine. The prototype and first production batch (marine numbers 1147–1156) of F2 floatplanes had angular vertical fin and rudder with an aerodynamic horn balance on the rudder. More than one vertical fin design was used by the F2 as shown by the photos. The final production batch (marine numbers 1974–1979) had rudders with the a aerodynamic balance within the outline of the fin and rudder that was very similar to the later F4 design. This made sense as the first production batch of the F4 was ordered before the last production batch of the F2. The entire first F2 production batch was assigned to Zeebrugge, but none of the final production batch were assigned to Zeebrugge. (Peter M. Grosz collection/STDB)

Above: Lübeck-Travemünde F2 (1147) was allocated to Zeebrugge on 27 August 1917. Powered by the 220 hp Mercedes D.IV straight-eight engine, the F2 was a massive type CHFT, meaning it had a wireless transmitter and receiver and a flexible machine gun for the observer. Its four-bladed propeller was needed to absorb the power from the slow-turning engine that was fitted with a reduction gear. This aircraft had a ventral fin extending below the fuselage to supplement the fairly small vertical fin about the fuselage.

Above: Lübeck-Travemünde F2 #1149 was allocated to Zeebrugge on 27 August 1917. Here it is shown in 1918 national insignia and the standard late-war Naval aircraft camouflage. (Peter M. Grosz collection/STDB)

Left: Enlargement of the Lübeck-Travemünde F2 from the photo of an F2 inspecting a sailing vessel. The rudder does not have a horn balance like the other F2 aircraft in the photos; instead it has an internal balance like the rudder of the F4. This is an F2 of the last production batch; the prototype and first production batch all had horn-balanced rudders. The ten aircraft of the first production batch were all assigned to Zeebrugge, but the last production batch was not. (Peter M. Grosz collection/STDB)

Lübeck-Travemünde F3

Above: Lübeck-Travemünde F3 fighter prototype marine number 844 had a 150 hp Benz Bz.III.

Lübeck-Travemünde F4

Above: Lübeck-Travemünde F4 before the marine number has been applied, so this may be the prototype. Evolved from the earlier F2 and designed for the same role, it was powered by the 200 hp Benz Bz.IV, which was less powerful but more reliable than the 220 hp Mercedes D.IV in the F2. Like the F2, the F4 was a massive type CHFT, meaning it had a wireless transmitter and receiver and a flexible machine gun for the observer. (Peter M. Grosz collection/STDB)

Above: Lübeck-Travemünde F4 marine number 1971 was the first prototype. (Peter M. Grosz collection/STDB)

LVG C.II Floatplane

Above: The LVG C.II floatplanes were built at Köslin, and this photo shows one during flotation tests at Köslin. The aircraft had a 160 hp Mercedes D.III engine; however, instead of a leading edge radiator normally used by C.II aircraft it used the earlier side radiators. Intended for Turkey, this aircraft had the work number K 81, the 'K' indicating it was built at Köslin. (Peter M. Grosz Collection/STDB)

Below: The struts attaching the floats to the airframe were much less robust than those of competing Friedrichshafen floatplanes, making the LVG C.II floatplanes unsuitable for use in the rough waters of the North Sea or Baltic. The Turkish insignia on the rudder shows this aircraft was intended for Turkey. The LVG company designation for these aircraft was D 9w, with the 'w' standing for '*Wasser*' (water). (Peter M. Grosz Collection/STDB)

Oertz F.B.1

Above: The Oertz F.B.1 was a pre-war design, being built in 1913. The F.B.1 was clearly a boat hull with flying surfaces.

Oertz F.B.2

Above: The Oertz F.B.2 was another pre-war design, being built in 1913. The F.B.2 was a refinement of the F.B.1.

Oertz F.B.3

Above: The Oertz F.B.3 flying boat(46) is towed by a boat out onto the water in the inner harbor for take-off. (M. Szigeti)

Below: Frontal view of the Oertz F.B.3 flying boat (46) on the mole at Zeebrugge. The similarities in construction between the hull and that of a yacht can be easily recognized.

Oertz W4

Above & Right: The Oertz W4 with national markings but before its Marine Number was applied. The excellent workmanship for which Oertz was famous is apparent. Two crewmen sat side by side in an open cockpit. Engine was a 160 hp Mercedes D.III.

Below: The Oertz W4 photographed in front of the Oertz factory.

Oertz W5

Above & Below: Oertz W5 Marine Number 276. The W5 followed the configuration of earlier Oertz designs but was larger and more powerful; the engine was a 240 hp Maybach Mb.IV. Five examples of the W5 were built, MN 276–280, making it the Oertz design built in greatest quantity.

Oertz W6

Above & Below: Oertz W6 Marine Number 281, with its tandem biplane design, was the most distinctive Oertz flying boat. Known as the *Flugschoner* (Flying Schooner), it was powered by two 240 hp Maybach Mb.IV engines. In original configuration above, the below photo shows it after auxiliary ailerons were added between the rear pair of wings.

Oertz W7

Above & Below: Oertz W7 marine number 474. Engine was a 150 hp Maybach Mb.III. (Peter M. Grosz collection/STDB)

Oertz W8

Above & Below: Oertz W8 Marine Number 1157. The engine was a 240 hp Maybach Mb.IV.

Rumpler 4B11

Above: Rumpler 4B11 Marine Number 49. The engine was a 100 hp Mercedes D.I.

Rumpler 4B12

Above: The Rumpler 4B12 was powered by a 150 hp Benz Bz.III. The Rumpler 4B series were developments of the 4A13 landplane. This unmarked aircraft may have been the prototype seaplane; it still uses the small rudder of its landplane ancestors while later seaplanes had a larger rudder due to the destabilizing effect of the floats.

Right: Rumpler 4B12 Marine Number 252 has yet another fin and rudder design.

The Rumpler 4B11, 4B12, and 4B13 were floatplane developments of the 4A13 landplane. They differed in their engines; the 4B11 used a 100 hp Mercedes, the 4B12 used a 150 hp Benz, and the 4B13 used a 160 hp Gnôme rotary.

Above: Rumpler 4B12 in flight. The rudder had a prominent horn balance.

Right: Rumpler 4B12 Marine Number 108.

Rumpler 4E

This Page: Like most Rumpler designs the Rumpler 4E was well thought out and constructed, with clean lines for a biplane flying boat. Regardless of its qualities, the German Navy favored floatplanes over flying boats due to the cold waters around Germany. Floatplanes kept the crew further away from the cold water and kept them dryer and warmer. It was assigned MN 47.

Rumpler 6B1

Above: The first prototype Rumpler 6B1, Marine#751, is here conducting flight trials at the Rumpler facility on the Müggelsee near Berlin in June 1916. The armament has not yet been installed. The engine was a 160 hp Mercedes D.III.

Above: Rumpler 6B1 Marine #1045 in flight. All 6B1 fighters except the three prototypes had the curved C.IV horizontal tail. The 6B1 had a large spinner.

Rumpler 6B2

Above: Rumpler 6B2 Marine #1189 was from the first 6B2 production batch. Its lack of propeller spinner was due to research to improve the Rumpler C.IV series that showed a streamlined nose without spinner had less drag. This made the 6B2 14 km/h faster than the similarly-powered 6B1, a significant improvement for such an apparently minor change. Here it is shown on its beaching gear at *Seeflugstation Puntisella* on July 9, 1918, in Austro-Hungarian service.

Above: Rumpler 6B2 Marine #1066 was from the 6B2 pre-production batch. From the rear it is difficult to tell the 6B1 from the 6B2. Although the 6B2 was designed to carry two machine guns, only about half the aircraft actually had two guns.

Sablatnig SF1

Above: This side view of the sole SF1 shows the very streamlined nose and rakish lines to advantage. The engine was a 160 hp Mercedes D.III.

Above: The rear quarter view of the SF1 emphasizes its streamlined vertical tail surfaces – and the struts bracing it. As a mechanical engineer and pilot, Sablatnig understood the need for streamlining but was unable to design a robust structure without excessive drag. In fact, the structural design was inadequate both in strength and aerodynamic cleanliness despite the well-shaped nose, as exemplified in the need for additional bracing struts and wires above the outer interplane struts.

Sablatnig SF2

Above: The first SF2, Marine #580, embodies the streamlined nose and rakish profile typical of early Sablatnig seaplanes, showing that Sablatnig understood the need for streamlining to achieve performance. The SF2 also exhibits the profusion of struts and bracing wires typical of Sablatnig seaplanes, revealing how difficult it was to design a sturdy yet low-drag airframe with the materials and design techniques of the time. Despite Sablatnig's advanced technical training, he was unable to advance the structural and aerodynamic state of the art for seaplanes. The result was the SF series had, at best, average performance and mediocre strength and durability. The SF2 had the same engine as the SF1 but a larger rudder.

Below: SF2 MN 799 was a production aircraft with larger vertical tail surfaces.

Sablatnig SF3

Above: The Sablatnig SF3 was compact for a two-seat fighter. However, the SF3 featured a profusion of struts and bracing wires that certainly created more drag than the cleverly-designed W12 and only one prototype was built. The engine was a 220 hp Benz Bz.IV.

Above: The Sablatnig SF3 prototype was finished in the standard late-war naval camouflage.

Sablatnig SF4 Biplane

Above & Below: The sole Sablatnig SF4 biplane prototype was Marine Number 900. The engine was a 150 hp Benz Bz.III. The SF4 competed with the Albatros W4, Brandenburg KDW, and Friedrichshafen FF43 but was not put into production due to is slow climb and fragile wing cellule.

Sablatnig SF4 Triplane

Above: Built to improve the rate of climb, apparently the performance of the SF4 triplane was no improvement over the SF4 biplane and, like the biplane, only one was built. It retained the 150 hp Benz of the biplane.

Sablatnig SF5

Above: Tiger-striped Marine Number 968 was the first SF5. The SF5 was closely based on the earlier SF2, although minor structural modifications were made to reduce weight. The most obvious change was the revised wing struts; the overhead bracing of the upper wing was replaced by an additional pair of struts attached to the front, outboard interplane strut. The 160 hp Mercedes used in the SF2 was in great demand for fighters, so the SF5 used the 150 hp Benz Bz.III.

Above: SF5 Marine Number 1230 from the fourth production batch. The SF6, also known as the Sablatnig B.I, was a landplane development of the SF5.

Sablatnig SF7

Above: The SF7 was a massive aircraft, looking more like a typical reconnaissance floatplane than a two-seat fighter. The competing W19 proved to be superior and was placed in production. Marine #1475 was the first of three prototypes. The engine was a 240 hp Maybach Mb.IV, the same engine used in the W19 and Friedrichshafen FF48 that were its competitors.

Above: The Sablatnig SF7 was in the W19 class and was powered by the same engine. The Friedrichshafen FF48 was the third competitor for this requirement. The I-struts appear to interfere with the crews' lateral field of view.

Sablatnig SF8

Above & Left: The SF8, a three-bay development of the SF5, was a dual-control floatplane trainer with 150 hp Benz Bz.III engine. Marine Number 2021 is shown below and is probably the aircraft above as well.

Right: SF8 Marine Number 2021 floatplane trainer; ailerons on all wings were connected by an actuating strut.

Union Floatplane Biplane

Above: Union floatplane biplane with 100 hp Mercedes D.I engine and side radiators. (Peter M. Grosz collection, STDB)

Below: Union floatplane biplane; one bought for the Navy was assigned MN 73. (Peter M. Grosz collection, STDB)

Zeppelin-Lindau Rs.I

Above: The metal Rs.I in front of the hangar in Seemoos in October 1915. (Airbus Group).

Above: The Rs.I on December 21st, 1915, before its final excursion into the water. It was wrecked by a storm before its first flight and was scrapped. (Airbus Group)

Zeppelin-Lindau Rs.II

Above: The metal Rs.II before its maiden flight on May 17th, 1916. The rudders are small and there are no fixed fins. (Airbus Group)

Above: On May 17th, 1916, the Rs.II was lowered into Lake Constance for its first test. (Airbus Group)

Above: The entire hull of the RS.II was overhauled and prepared for the tandem engines. The aerodynamics were improved, rudders and fin enlarged, and it was painted grey-green. (Airbus Group)

Above: Front view of the RS.II in is second configuration after overhaul and for tandem engines installed. (Airbus Group)

Below: Rs.II taxiing on Lake Constance on November 11th, 1916. Note the larger fins and rudders. (Airbus Group)

Zeppelin-Lindau Rs.III

Above: The all-metal Rs.III monoplane flying boat ready for launch.

Above: In March 1918, naval pilots brought the Rs.III to the SVK in Warnemünde, where observers marveled at her. (Airbus Group)

Below: The Rs.III before she was brought into the hangar. In the background are a few FF 49c aircraft.

Zeppelin-Lindau Rs.IV

Above: On October 12th, 1918, the all-metal Rs.IV took off on its maiden flight. (Airbus Group)

Below: Close-up of the metal Rs.IV. (Airbus Group)

Zeppelin-Lindau Cs.I

Above & Right: The ZWL Cs.I MN 8502 with ear radiators and spinner. The engine was a 195 hp Benz Bz.IIIb V-8. (PM Grosz collection/STDB)

Below: Closeup of the ZWL Cs.I MN 8503 with production nose radiator. (PM Grosz collection/STDB)

Zeppelin-Lindau Gs.I

Right: Model of the ZLW Gs.I in its original, military configuration. The engines were two 260 hp Maybach Mb.IVa. (PM Grosz collection/STDB)

Above, Below, & Right: The ZLW Gs.I in Swiss civilian service postwar. The passenger cabin added to convert the Gs.I to civil service is prominent. The letters 'CH' indicate Swiss civil registration and '8' is on the fuselage side. (PM Grosz collection/STDB)

Zeppelin-Staaken L

Above: The Staaken L was essentially an R.VI on floats with additional equipment to suit is role as a long-range maritime reconnaissance airplane and bomber. A fixed fin and third rudder were added to improve stability and control, and the ailerons were balanced to reduce control forces. Assigned marine number 1432, it was modified a number of times in light of test-flight results; here it is seen in the 4th of five configurations. (Peter M. Grosz collection/STDB)

Below: The Staaken L 1432 in the 3rd of five configurations. In its first two configurations the 1432 did not have the central fin and rudder. Power was four 260 hp Mercedes D.IVa engines. (Peter M. Grosz collection/STDB)

Zeppelin-Staaken 8301 & 8303

Above: The Staaken 8301 and 8303 were developments of the earlier Staaken L. The Staaken L had been successful enough that the Navy ordered four aircraft that were developments of it; the 8301 (above) and 8302, then the 8303 and 8304 that had slight improvements. The most significant improvement was raising the fuselage above the lower wing to get the airframe and tail further out of the water. The nose was also slightly redesigned to give all-around visibility for its maritime reconnaissance role. (Peter M. Grosz collection/STDB)

Below: The Staaken 8301 was completed as a land-plane to facilitate initial flight testing before its floats were fitted. Like the 'L', power was from four 260 hp Mercedes D.IVa engines. (Peter M. Grosz collection/STDB)

Made in United States
Cleveland, OH
05 November 2026

47673399R00103